Ahlam Mohamad

Managing the Potential of Modularization and Standardization of MEP Systems in Buildings

Guidelines for improvement based on lean principles

Karlsruher Reihe Technologie und Management im Baubetrieb
Karlsruher Institut für Technologie (KIT)
Institut für Technologie und Management im Baubetrieb

Prof. Dr.-Ing. Dipl.-Kfm. Shervin Haghsheno (Hrsg.)
Prof. Dr.-Ing. Sascha Gentes (Hrsg.)

Heft 72

Das Institut für Technologie und Management im Baubetrieb (TMB) befasst sich in Forschung und Lehre mit dem gesamten Bereich des Baubetriebs von der Maschinen- und Verfahrenstechnik bis hin zum Management der Projekte, Facilities und Unternehmen.
Weitere Informationen und Kontakte unter www.tmb.kit.edu

Eine Übersicht der Forschungsberichte finden Sie am Ende des Buches.

Managing the Potential of Modularization and Standardization of MEP Systems in Buildings

Guidelines for improvement based on lean principles

by
Ahlam Mohamad

KIT Scientific Publishing

Karlsruher Institut für Technologie
Institut für Technologie und Management im Baubetrieb

Managing the Potential of Modularization and Standardization of MEP Systems in Buildings – Guidelines for improvement based on lean principles

Zur Erlangung des akademischen Grades einer Doktor-Ingenieurin von der KIT-Fakultät für Bauingenieur-, Geo- und Umweltwissenschaften des Karlsruher Instituts für Technologie (KIT) genehmigte Dissertation

von M.Sc. Ahlam Mohamad aus Syrien

Tag der mündlichen Prüfung: 23. November 2015
Referent: Prof. Dr. Ing. Fritz Gehbauer, M.S.
Korreferent: Prof. Dr. Ing. Sascha Gentes

Impressum

KIT Scientific Publishing

Karlsruher Institut für Technologie (KIT)
KIT Scientific Publishing
Straße am Forum 2
D-76131 Karlsruhe

KIT Scientific Publishing is a registered trademark of Karlsruhe Institute of Technology. Reprint using the book cover is not allowed.

www.ksp.kit.edu

Print on Demand 2019 – Gedruckt auf FSC-zertifiziertem Papier

ISSN 2363-8222
ISBN 978-3-7315-0556-3
DOI 10.5445/KSP/1000057591

Acknowledgements

I would like to express my special appreciation and thanks to my advisor Professor Dr.-Ing. Fritz Gehbauer for providing me with the possibility of working and pursuing my research at institute for technology and management in construction and for his excellent personal and scientific comportment. I would also like to express my special thanks to my second reviewer Professor Dr.-Ing. Sascha Gentes for his scientific comments and help.

Apart from my advisor and second reviewer, I would like to thank the rest of my thesis committee: Professor Dr.-Ing. Kunibert Lennerts and Professor Dr.-Ing. Petra von Both for accepting the request to serve as reviewers in the examination committee and for taking the time to give insightful comments, to pose hard questions, and to provide helpful advice.

Special thanks go to: Volkmar Hovestadt, Uwe Batzler, Herrmann and Bernd Knobel who helped me to get information from real projects to be able to complete my dissertation. I would also like to express my thanks to my colleagues at Institute for Technology and Management in Construction. In particular, I would like to thank: Gernot Hickethier, Annett Schöttle, Markus Reinhardt, Heiner Schlick, Harald Schneider and Tobias Bregenhorn.

My special thank goes to the Syrian Ministry of Education and Karlsruhe House of Young Scientists (KHYS) for financing my research.

Lastly, but most importantly, infinitely many gratitude to my husband for his love and support and to my daughter and my son for just being there for me. You have helped me to stay calm and positive in stressful times and gave me the courage to pursue this path until the end.

Karlsruhe, Januar 2016

Ahlam Mohamad

Abstract

Modularization and standardization (M&S) of MEP (mechanical, electrical, and plumbing) systems improve end customer value according to lean concepts and principles. However, the implementation of M&S is challenged. This research includes three main parts:

First, the advantages of M&S of MEP systems in improving the construction process and increasing flexibility against design changes are explained through literature review and a case study. By showing how M&S reduce the variety of the components (a cause for variability in the construction process), the case study explains how the construction process can be improved. In addition, depending on the interviews and literature review, the advantages of increasing flexibility against design change are also explained.

Second, the challenges of the implementation of M&S were investigated through reviewing documents and conducting interviews with different project partners, like designer, construction manager, facility manager, subcontractor, and installer. The investigation shows that the challenges are distributed across design and construction phases, and that the organizational concepts, like types of contracts, are essential to achieve a successful implementation.

Third, a guidelines model is developed, based on the executed case studies, the interviews, and literature of M&S. Suggestions to manage the implementation of M&S of MEP systems are made depending on lean concepts and tools.

The case studies and interviews were used to understand the phenomena of M&S of MEP systems, and they facilitated building a theory about their implementation. The focus was on different project phases to explain challenges of implementation, highlighting several typical elements of a standard practice "traditional sequential management system". The results show how implementation of modularization and standardization, to increase the value of the product, fit into the lean project delivery system.

Table of contents

List of figures

List of tables

Abbreviation

AEC	Architecture, Engineering and Construction
CI	Coupling Index
CI–R	Coupling Index -Receive
DfX	Design for X
DFV	Design for variety
DSM	Design Structure Matrix
GEMS	Geometry, Energy, Material, or Signal
GVI	Generational Variety Index
HoQ	Hous of Quality
LPD	Lean Product Development
LPS	Last Planner System
M&S	Modularization and Standardization
MEP	Mechanical, Electrical and Plumbing systems
M&E	Mechanical and Electrical
MID	Modularity in Design
MIP	Modularity in Production
MIU	Modularity in use
M&S	Modularization and Standardization
MFD	Modular Function Deployment
PDCA	Plan, Do, Check, Act cycle
QFD	Quality function deployment
VD	virtual Design

1 Introduction

1.1 M&S of MEP systems

Modularization and standardization (M&S) have been used successfully very early in different industries, and the results are large savings during the life cycle of products such as computers, ships and many electrical and technical products.

Mechanical, electrical, and plumbing (MEP) systems in the construction industry are complex systems (products) and they contribute to 40% to 60% of the construction costs in industrial buildings (Khanzode 2011). Improvement of the architecture of these systems will lead to high savings for the construction industry. These Improvements could be achieved through applying M&S concepts during the design process.

M&S of MEP systems deal with the architecture of these systems, which depends on geometrical, structural and utilization aspects of the building. They can improve the design and construction process by reducing the variety of components, and allowing flexibility where a variety of utilizations can be created using standardized components or structures of MEP systems. However, how can M&S of MEP systems be achieved? Developing a design methodology is an important aspect, but it cannot achieve this goal alone. Managing the developing process is essential because of many interdependencies in the course of applying M&S of MEP systems, and because of a large number of participants whose knowledge needs to be managed efficiently and effectively.

Figure 1.1 shows the implementation of M&S observed in the practice in different phases of project planning and the benefits gained from them:

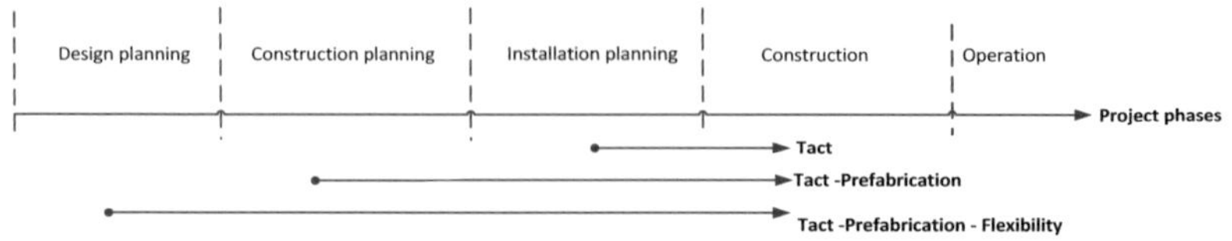

Figure 1.1: Implementation of M&S of MEPs in different phases of the project planning and the gained benefits

It can be seen from Figure 1.1 that the type and extent of the advantages depend greatly on the phase of the implementation. Although the implementation in design allows gaining more benefits (tact, prefabrication, flexibility), many challenges occurred during this implementation, which will be described later through this research. Also, a task of this research work is to discuss possibilities to reduce or eliminate these challenges as a way to increase the possibility to realize the potentials of M&S of MEP systems.

This research shows how using lean concepts and tools helps managing the design process to achieve M&S of MEP systems in design and construction. In current practice, design and construction are separated legally to get low-price offers. The work of the designer is regulated according to HOAI (Honorarordnung für Architekten und Ingenioure), and the separated departments are the general rules. These characteristics of the management system make the learning process very slow, especially when knowledge of different specialists, including construction teams, is needed to improve decision making and feasibility learning, which are all elements of successful implementation of M&S. The executed case studies and interviews show how current thinking and management practice contribute to hinder the effective and efficient implementation of M&S.

M&S relate to a product development process, which is a production process (Ballard 2000a). The research claims that integration of construction firms during design is a management aspect that helps achieve M&S. A further aspect of managing design process to achieve M&S of MEP systems is managing adjustment processes during design to improve M&S efficiently.

Understanding the characteristics of M&S processes of MEP systems is an important factor to define management patterns. In addition, the inspection of challenges and occasional successful factors propose a ground to inspect possibilities to reduce or eliminate them. These possibilities will be analyzed from the lean thinking (and tools) perspective.

In this research, developing a guideline-model to implement and manage M&S of MEP systems will facilitate managing workflow and communication process during the implementation. The research shows how implementation of M&S to increase the value of the product (and end customer) fits into the lean project delivery system, and it includes implications like teams, culture, organizations, and other aspects.

1.2 Goal

The goal of this research is to develop guidelines to achieve modularization and standardization (M&S) of MEP systems in design to increase the value of a product. The guidelines aim to manage the potential of M&S, and it strives to improve the performance of implementation.

1.3 Research questions

The research questions to achieve the goal are:

1. How to apply M&S concepts on MEP systems?
2. What are the challenges faced in achieving M&S of MEP systems?

3. How to manage the potential of M&S from the lean perspective?

The research questions focus on two main points: First, inspecting the adaptability of M&S on MEP systems as a basis to describe their characteristics. Second, the impacts of the functional, "traditional" management system on this adaptability; these impacts can be expressed in terms of efficiency (when the focus is on types of waste) and effectiveness (when the focus is on realizing modularized and standardized architecture of MEP systems).

1.4 Methodology

The methodology to answer the research questions includes inspecting the challenges of M&S along project phases. This is made through inspecting the implementation during design and construction processes in completed and ongoing projects. Interviewing different participants, namely, designers, users, construction managers, and installers has introduced a wide knowledge about current implementations and challenges. Case studies were used to understand the phenomena of modularization and standardization and then to build a theory (Meredith 1998).

First, a design methodology for applying M&S of MEP systems with connection to concepts in the literature is introduced though interviews and discussions with the designer who co-developed the design methodology. Second, one completed project, where M&S had been implemented, was inspected through interviews with the owner and installer. Third, the results of the implementation of M&S on the construction site were inspected on an ongoing project through interview with the construction manager and designer. Fourth, observations during the design process in an ongoing project were used to define real challenges of possible implementation of M&S of MEP systems. Fifth, interviews were made with design management teams to capture types of waste during the implementation in design .Then, a guideline to manage implementation, depending on the obtained knowledge from the case studies is developed, literature review, and the deep analysis. Figure 1.2 shows the flow of the research work:

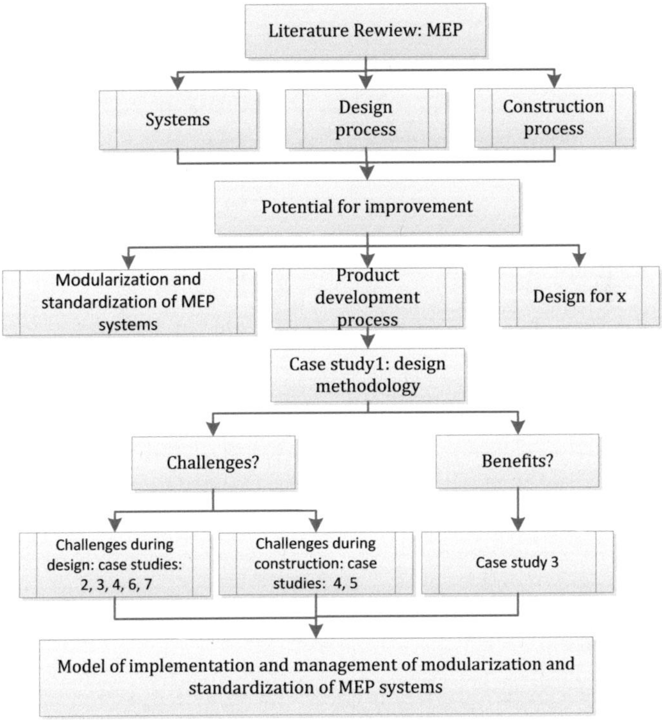

Figure 1.2: Methodology

The research methodology can be described as follows:

1. Analyzing M&S processes and methods
2. Inspecting the adaptability of implementation of M&S on MEP systems
3. Defining challenges caused by adaptability

4. Developing a model of management system using lean tools and methods, depending on:
   - Characterizing the process of applying M&S to the design process
   - Integrating construction factors, especially manufacturing quality and warranty claims, while applying M&S processes during the design process
   - Measuring the reliability of implementation using developed indicators.

Our methods in this research are:

1. Non-participant observation in the design process
2. Qualitative interviews with different projects' participants
3. Analysis of projects' materials.

Seven case studies were performed. The case studies were distributed during the design and construction phases of different projects because it was not possible to inspect the research aspects in only one project.

1.5 Outline of dissertation

Chapter 2 describes the planning process of MEP systems where high interdependencies exist with building systems, and the obstacles in the construction process of MEP systems. Chapter 3 discusses the product development process and methods under which M&S can be classified. Chapter 4 describes M&S and their benefits and methods. Chapter 5 describes aspects of design management and lean thinking in design. The case studies are introduced in Chapter 6. Guideline to manage implementation of M&S is introduced in Chapter. Finally, conclusion and future research are introduced in Chapter 8.

2 MEP Systems

MEP systems on construction projects account for 20% to 40% of the total project cost, whereas the complexity of these systems has increased over the years (Khanzode 2011). Therefore, it is essential to consider MEPs' design, fabrication, and installation process of these systems (Tatum and Korman 2000).

2.1 Planning process of MEP systems

2.1.1 Coordination process of MEP systems

The MEP coordination process bridges the gap between design, fabrication, installation, and operation of these systems (Khanzode 2011), and it is needed to identify the location and configuration of these systems (HVAC, plumbing, fire protection, etc.) (Tatum and Korman 2000). However, fragmented responsibility and required knowledge of design, installation, and operation lead to challenges for the coordination process, for it specialty- or trade contractors are typically responsible (e.g., Tatum and Korman 2000).

The coordination of MEP systems is done by the specialty contractors, where the engineer or the designer prepare diagrammatic drawings but with no detailed layout and installation instruction (e.g., Tatum and Korman 2000). Current interviews conducted within the framework of this research confirm the previous statement.

The process of MEP systems' coordination involves locating of components and branches of all systems in congested spaces, considering design, construction, and maintenance criteria, like spatial (avoiding interferences), functionality within a system (flow or gravity drainage), adjacency or segregation, system installation (layout dimensions, space and access for installation productivity), and testing (ability to isolate) (Tatum and Korman 1999).

According to Tatum and Korman (2000), the knowledge required for the coordination process consists of: Design knowledge, construction knowledge, and knowledge of the facility life cycle.

One challenge in the coordination process is the existence of limited spaces for MEP systems; this challenges the design and construction processes of these systems (Tatum and Korman 1999). Typically, the MEP systems are not designed with details because of limited available time or reduced fees for the designer, so contractors or installer are responsible for detailing the systems. Therefore, the coordination process will be more challenging. Further, a major challenge is capturing distributed knowledge of the different types of MEP systems and tailoring the software to coordinate special needs of MEP systems (Tatum and Korman 1999).

2.1.2 Interdependencies in MEP systems

There are many types of interdependencies in designing products. Schmidt et al. (Eppinger and Browning 2012) captured three types of dependencies in the design process of building: 1) structural, 2), spatial, and 3) services to define possibility of modularization for adaptability. In their research, they found that dependencies existed between the layers of the building (skin, services, space plan, stuff, space, and site) which had been defined before by Brand (1995).

Khanzode (2011) describes the challenges in the coordination process of MEP systems' design, highlighting the reciprocal nature in the coordination process. Figure 2.1 represents an example of the reciprocal dependency in the coordination process:

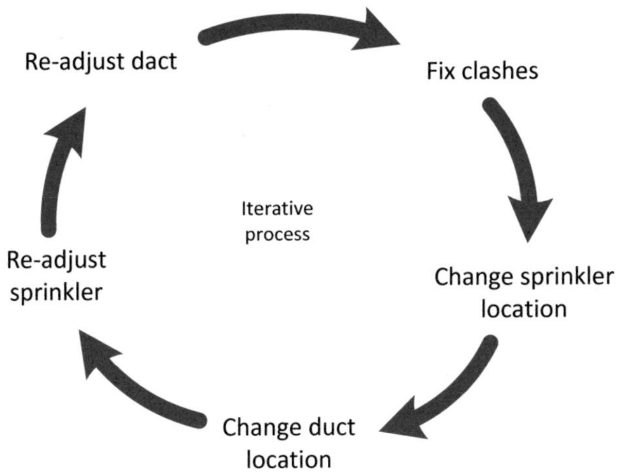

Figure 2.1: Example: Reciprocal dependency in MEP's coordination process (Khanzode 2011)

This reciprocal dependency can be solved through back and forth negotiations between the trade contractors (Khanzode 2011). Khanzode (2011) studied the required level of detail when applying virtual design (VD) methodologies to the MEP coordination process, as can be seen in Figure 2.1 which also explains the dependencies between MEP systems and other building systems:

Table 2.1: Levels of details above 1/4" diameter in the various models need to be addressed as part of the technical logistics (Khanzode 2011)

Model	Level of Detail	Benefit in MEP Coordination Process
Architectural model	Wall with attributes of thickness and height	Required for routing main utilities, locating VAV boxes, identifying priority wall framing, wall penetrations, fire stopping
	Hard ceilings and soffits	Required for identifying HVAC diffuser locations, electrical fixture locations, routing of utilities
	Suspended acoustical ceilings	Required for identifying HVAC diffuser locations, electrical fixture locations, routing of utilities
	Casework with correct fixture locations	Useful for identifying the location of plumbing utilities rough-in in the walls
	Exterior wall / store-front	Useful for identifying the locations of rain water leaders
	Shafts, wall chases	Required for identifying the correct locations of plumbing vents, HVAC shafts
Structural model	Foundations, grade beams	Required for coordination of under-ground utilities like electrical, plumbing
	Beams and columns	Required for coordinating above ceiling MEP / FP utilities
	Braces and gusset plates	Required for coordination the routing of MEP piping
	Miscellaneous support steel like exam light supports for medical facilities or Unistrut, etc.	Required for the correct routing of MEP utilities correctly
	External wall framing connections (like con-nection between steel and GFRC panels)	Required to coordinate the plumbing rain water leaders
Mechanical model	Medium-pressure ducts	Required for coordination and routing of other trades as well as prefabrication
	Low-pressure ducts	Required for coordination and routing of other trades as well as prefabrication
	Shaft locations	Required for coordination and routing of other trades and for locating smoke dampers, etc.

	VAV boxes	Required for prefabrication purposes, coordination with HVAC heating hot water piping
	Fire smoke dampers	Useful for coordination, especially if walls are also provided in the model
	Flex ducts	Useful for showing how low-pressure duct connects to the diffusers
	Diffuser locations	Useful for coordination of finish utilities with the other fixtures in a room (like electrical fixtures, etc.)
	Hangers and seismic bracing	Required for coordination and routing of other trades and for inserting the deck correctly before installation begins
	HVAC Piping to the VAV boxes	Main lines are required for coordinating with other trades. Also required if they will be prefabricated. Connections to VAV boxes can be left for field routing.
	Rooftop equipment	Useful for coordinating with other trades (can be drawn as a 3D block)
Electrical model	Branch and feeder conduits	Required for coordination with other trades and for prefabrication
	All underground conduits	Required for underground MEP / FP coordination
	Junction boxes	Required for coordination with other trades
	All lighting fixtures	Required for coordination with other trades and finish utilities like ceiling grid, sprinkler heads, HVAC diffusers, specialty lighting
	All lighting supports in case of specialty lighting	Required for routing and coordination of other trades
	All the cable trays and other supports	Required for coordination with other trades
	Bundles of cables or wiring	Useful for coordination and prefabrication
	Outlets and switch locations in rooms	Useful for prefabrication but not required for coordination with other trades (typically can model in 2D)
	Hangers	Required for coordination with other trades and for inserting the deck.

	Equipment panels	Useful for coordination with wall framing to determine backing, etc.
	Electrical rooms	Useful for coordination with wall framing and other trades
Plumbing model	Plumbing fixtures	Required for coordination with other MEP trades and casework
	Graded cast iron pipe lines	Required for coordination with other trades and for prefabrication
	Underground storm and sewer pipes	Required for underground utilities coordination and for prefabrication
	All major waste and vent lines	Required for coordination with other trades and with architectural walls, shafts, and for prefabrication
	Cold and hot water piping	Required for coordination with other trades and for prefabrication
	Hangers and seismic bracing details	Required for coordination with other trades and for inserting before installation and for prefabrication
	All boilers and other equipment	Useful for coordination (can be drawn as a 3D block)
	Specialty piping (like medical gas piping) and specialty equipment	Required for coordination with other trades and for prefabrication
Sprinkler model	Sprinkler mains and branches	Required for coordination with other trades and for prefabrication
	Sprinkler head drops	Useful for coordination with finish utilities like electrical lighting, diffusers, etc.
	Smaller sprinkler pipe	Required if hard pipe is used, useful if the newer type of flex pipe is used

Also, the architecture of the MEP systems, which can be described through the configuration of these systems and their dependencies, has a great impact on the extent to which the building is flexible against changes in the future.

2.2 Construction process of MEP systems

There are difficulties and challenges in managing the construction process of MEP systems because of its features. Three main features of the behavior of the trade crews could be observed: (1) irregular production patterns and rates, including interference between crews, (2) interruptions in the work applied to spaces and work-packages, and (3) re-entrant flow patterns (Brodetskaia et al. 2011). Brodetskaia et al. (2011) have modeled the workflow of these works to help manage them by adapting suitable management policies according to the characteristic of the workflow of these works.

There are many types of waste and non-value generated activities in the construction processes of MEP systems. Figure 2.2 illustrates some types of waste and the instability in the workflow resulted through a field study made by Brodetskaia et al. (2011):

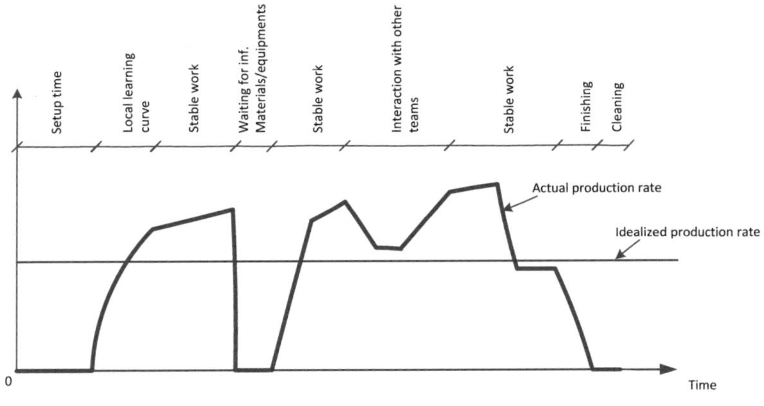

Figure 2.2: Typical fluctuations and interruptions to value generation through execution of a finishing activity (adapted from Brodetskaia et al. 2011)

Seppänen (2009), in three large case studies in Finland, found that many more problems are caused by MEP and interior works than foundation, structural- and facade works. The interruptions in the workflow are unpredictable and have uncertain durations (Brodetskaia et al. 2011), and the variations are caused because of uncertainties in supply chains, variations in work quantities, client changes, and lack of predictability of the production capacity of sub-contracting trades (Brodetskaia et al. 2012). According to Brodetskaia et al. (2012), in the interior and finishing works, the supply chains for materials are complex, varied, and potentially unreliable. Productivity rates of these works are highly variable (Radosavljević and Horner 2002). Court et al. (2009b) state that the productivity is poor on the construction site of mechanical and electrical (M&E) construction in the UK, thus causing delays, and that improving the efficiency of M&E is needed.

Also, according to interviews conducted for this research with many practitioners in Germany, MEP's workflows on the construction site contain different types of waste which then cause great delays on the construction project.

As a result of analyzing construction and coordination process of MEP systems: the construction process includes many types of waste, some of which return greatly to the architecture of these systems which describes variety and distribution of work quantities at the work-stations. Therefore, many types of waste return to the development process of these systems. The ability to influence the development process with the goal of developing architectures that can be simply managed on the construction site could be an effective strategy to reduce many types of waste during the construction phase. Also, the architecture of the MEP systems affects the behavior of the building regarding its flexibility against design changes.

In the following section, the product development process will be studied to understand the theory and background of the product development process of MEP systems.

3 Product development process

Product architecture is the "scheme by which the function of a product is allocated to physical components" (Ulrich 1995). More precisely, product architecture can be defined as: (1) the arrangement of functional elements; (2) the mapping from functional elements to physical components; and (3) the specification of the interfaces among interacting physical components (Ulrich 1995). A family architecture means that different products of the family have common mapping between functions and structure, common interactions among components, and common arrangements between elements (Martin and Ishii 2002).

Ulrich (1995) defined potential linkage between product architecture and five important managerial aspects: (1) product change; (2) product variety; (3) component standardization; (4) product performance; and (5) product development management.

A robust product design has the purpose of making the design insensitive to uncontrollable factors. The extent to which it can be made is the goal of robustness, while the ease of change is the goal of customization (Jiao and Tseng 2004).

3.1 Product development process and system engineering

The development process can be divided into systems, methods, and processes which link targets, information (knowledge), and activities (Albers and Meboldt 2007). Product development can be described, according to Albers and Meboldt (2007), as "the transfer from a system of objectives, being still vague at the beginning of the product development, to a concrete object system. i.e., the core activity of the product development is the continuous expansion and specification of a system of objectives, the creation of an efficient operation system and therefore the successful realization into an object system: the product". Figure 3.1 explains the previous definition of product development process.

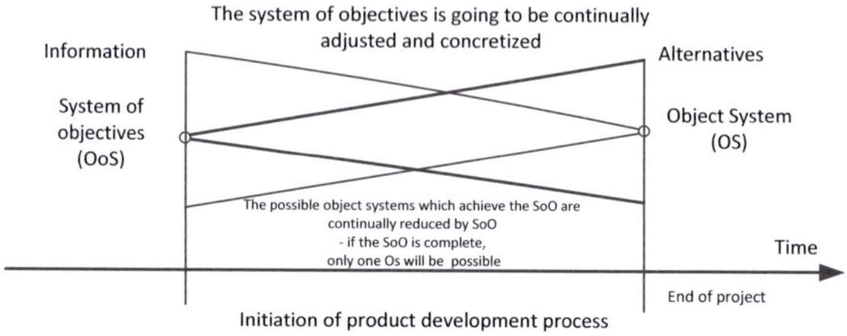

Figure 3.1: System of objectives, object system in the product development process (adapted from Albers and Meboldt 2007)

In the integrated product development, the product life cycle is described by means of systems engineering where the impacts of all systems on the product are considered (Albers and Meboldt 2007). Figure 3.2 shows the interactions between the different life cycle stages of the product.

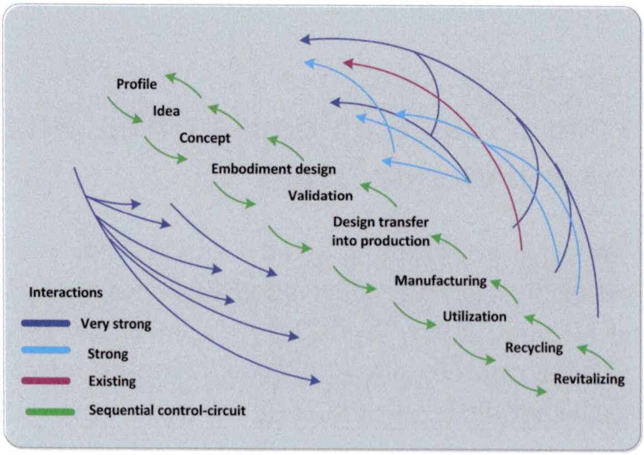

Figure 3.2: Stages of product life cycle (Albers and Meboldt 2007)

The challenge in system engineering is coordinating the interactions of the different stages and establishing a standard process for that (Cooper 1994).

A product development process is a problem solving process. A problem can be defined through three elements (Albers et al. 2005), as can be seen in Figure 3.3. Albers et al. (2005) define the problem as follows: "A problem is a deviation between the arbitrarily little known initial state (Actual State) and the desired arbitrarily vague final state (Target State), linked with the partially unknown path from the Actual to the Target State." (Albers et al. 2005).

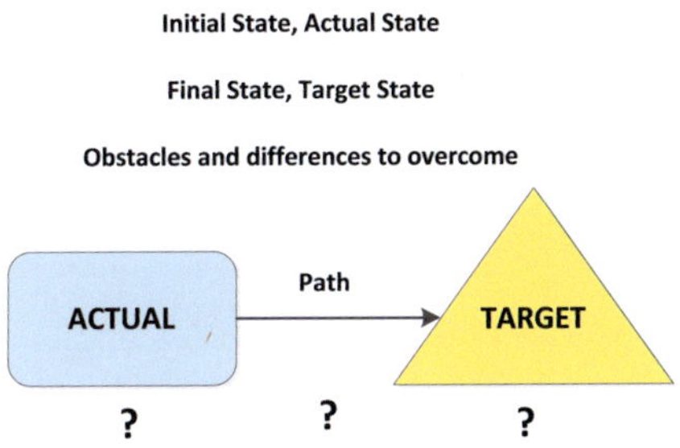

Figure 3.3: Component of a problem (Albers et al. 2005)

Problem solving in the product development process has two dimensions: The dimension of the life cycle and the problem solving of the single stages (Albers and Meboldt 2007).

One tool of problem solving in the product development process is the TOTE-schema (test-operate-test-exit) which aims to achieve the target state through making changes of the actual state and can be considered as a closed loop (Schregenberger 1980). Another tool of problem solving, used in lean product development and lean design, is A3 on the basis of the cycle (Plan-Do-Check-Act) developed by Deming. This tool will be discussed in Chapter 5.

3.2 Design for X (DFX)

Approaches of Design for X (DFX) provide qualitative design guidelines for a specific stage in a product lifecycle (e.g., Design for Manufacturing), or a specific virtue (e.g., Design for Environment), where X stands for a particular life phase or a virtue that the product should possess (Dombrowski et al. 2014, Holt and Barnes 2010). Table 3.1 shows important DFX methods (Holt and Barnes 2010):

Table 3.1: Examples of DFX_{virtue} and $DFX_{lifephase}$ techniques

DFX_{virtue}	**$DFX_{lifephase}$**
Design for environment	Design for manufacture and assembly
Design for quality	Design for end-of-life
Design for maintainability	Design for disassembly
Design for reliability	Design for recycling
Design for cost	Design for supply chain
Affective design	
Inclusive design	

Dombrowski et al. (2014) distinguish between product properties (like dimensions) and product characteristics (like reliability or maintainability) to develop a model to integrate DFX with lean design. They view product properties as equivalent to design view according to Lean Design and product characteristics describe the contribution of the product design to customer value (Value View) and the effect on lifecycle processes (Waste View), as shown in Figure 3.4:

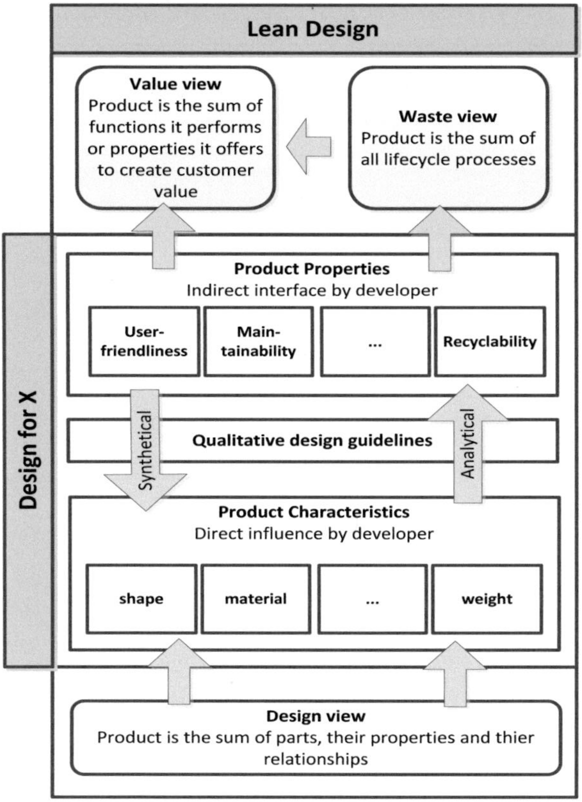

Figure 3.4: Integration model for DFX and lean design (Dombrowski et al. 2014)

The following matrix in Table 3.2 represents a qualitative design guideline developed by Dombrowski et al. (2014), explaining that developing a product with a specific characteristic to achieve certain properties (one or more) should consider the trade-offs to other product properties.

In this research, we will discuss a used method of design for variety applied on MEP systems, therefore we analyze in the next section design for variety and its methods.

Table 3.2: Analysis of qualitative design guidelines (Dombrowski et al. 2014)

	Product Characteristic	Product Property												Number of mentions		
		DfX lifephase				DfX virtue										
	Qualitative Design Guideline	Manufacturing	Assembly	Service	Recycling	Remanufacturing	Logistics	Environment	Quality	Maintainability	Reliability	Safety	User-friendliness	Total	Lifephase	Virtue
1	Minimize the number of parts	X	X	X	X		X	X	X	X	X			9	4	5
2	Develop a modular design	X	X	X		X	X	X	X	X				8	3	5
3	Avoid separate fasteners	X	X	X	X	X		X		X	X			8	4	4
4	Use standard components	X	X	X			X	X			X			6	3	3
5	Sharp edges, corners, or protrusions that could cause injury shall be avoided			X	X		X			X		X	X	6	2	4
6	Provide easy access for locating surfaces, symmetrical parts or exaggerate asymmetry		X	X	X					X		X		5	3	2
7	Design parts to be multi-useable	X			X	X	X				X			5	2	3
8	Minimize the needs for special tools	X	X	X	X					X				5	4	1
9	Use of proven components		X						X		X		X	4	1	3

10	Making the design insensitive to all uncontrollable source of variation								x	X	X	X		**4**	**0**	**4**
11	Minimize the number of design variants	X	X											**3**	**1**	**2**
12	Provide simple handling and trans-portation	X	X							X				**3**	**1**	**2**
13	Design parts that cannot be installed incorrectly (Poka Yoke)	X	X						X					**3**	**1**	**2**
14	Avoid hazardous and otherwise environmentally harmful materials				X			X				X		**3**	**1**	**2**
...	...													**...**	**...**	**...**
96	Make the controls and their functions obvious, provide direct feedback from the product												X	**1**	**0**	**1**
	Number of qualitative design guide-lines provided	13	35	11	13	6	22	14	15	16	19	8	9			

3.2.1 Design for variety

Variety is one of the elements of "Lean production" which has been identified as a successful factor in automobile manufacturing (Womack and Jones 1990). Product variety is "the diversity of products that a production system provides to the marketplace" (Ulrich 1995).

Design for variety (DFV) is "a series of structured methodologies to help design teams reduce the impact of variety on the life cycle costs of a product" (Martin and Ishii 2002). Design for variety aims to help the engineers in creating design that builds on current design efforts and then reduce development costs (Martin and Ishii 2002).

Martin and Ishii (2002) define two types of variety that should be considered in developing the architecture of the product: Spatial variety and generational variety. Spatial variety refers to variety within the product being designed, whereas generational variety refers to variety across generations. The drivers of the generational variety depend on uncontrollable factors (external factors), such as customer needs, reliability requirements, or reduced prices, as can be seen in Table 3.3:

Table 3.3: External drivers of generational change (Martin and Ishii 2002)

Customer requirements
Changing performance needs (including size, style, weight, etc.)
New environmental constraints (temperature, humidity, vibration, etc.)
New functions (due to new markets or new enabling technologies)
Reliability improvements
Reduced prices (cost reductions required)
Reduce amount of material
Change material type
Remove redundant components
Reduce assembly time

Use lower cost technology Reduce serviceability requirements Reduce serviceability time Improve component manufacturing process Regulations, standards, and so on Changing government/industry regulations or standards Competitor introduction of improved product (higher quality or lower price) Obsolescence of parts

Ulrich (1995) recognizes two types of change: Change within the life of a particular artifact and change across generations of the product. Change within the life of the particular artifact includes: Upgrade, add-ons, adaption, wear, consumption, and flexibility in use. Jensen et al. (2012) state that the customer preferences are the main driver of changing the modules that define the product platform.

Martin and Ishii (2002) developed two indices that contribute to design for variety: The generational variety index (GVI) that is caused by the external drivers, and the coupling index (CI) that is caused by internal drivers which come from the coupling between the product components and measured by the CI–R (Coupling Index -Receive). The internal and external drivers cause changing of a component. Figure 3.5 shows the drivers of component's changes:

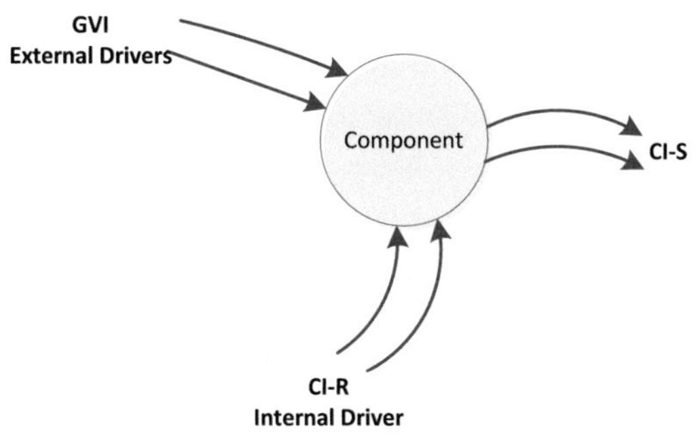

Figure 3.5: Illustrations of drivers of component change (Martin and Ishii 2002)

Two components are considered coupled "if a change made to one of the components can require the other component to change" (Ulrich 1995).

Martin and Ishii (2002) used modularization and standardization in the product architecture to achieve design for variety (M&S will be discussed in detail in Chapter 4). They defined the two concepts as follows:

1. Standardized (GVI and CI–R related):

   - Fully standardized: It is expected that the component will not change across generations. This implies that the GVI and CI–R are equal to zero.
   - Partially standardized: The component is expected to require minor changes across generations. The higher the GVI and CI–R, the less standardized is the component.

2. Modularized (CI–S related)

   - Fully modularized: The geometry, energy, material, or signal (GEMS) of the component can be changed to meet expected customer requirements without requiring other components to change. This implies that the CI–S of the component is zero.
   - Partially modularized: Changes in the GEMS of the component may require changes in other components. The higher the CI–S, the more changes expected, and thus the component is considered less modular.

Within the process of design for variety, design teams should make decisions about how to arrange the mapping between functions and physical components, and how to define interfaces (Martin and Ishii 2002). Overdesign is a method to reduce the sensitivity of components to changes, but it may increase cost of material (Martin and Ishii 2002).

Variety increases customer value if the functionality of the product changes, where functionality in this context means "any attribute of the product from which the user derives a benefit" (Ulrich 1995). The challenge in product variety is to create the desired product variety economically, which is frequently credited to manufacturing flexibility (Ulrich 1995). Upton (1994) defined flexibility as "the ability to change or adapt with little effort, time, or penalty".

Both, the flexibility of the factory production process equipment and the product architecture interact to contribute to the ability to economically create product variety (Ulrich 1995). Three aspects of cost justification should be considered when analyzing flexibility (or customizability) (Jiao and Tseng 2004): Utility, design changes, and process variation. Jiao and Tseng (2004) define two sources of customization: Design change and process variation, where design change relates to product variety results in process variation (process variety) that represents the impact of design change on the production process. Developing platforms for product and process aim to achieve mass production efficiency by using common structures of product and process to generate product variety (Jiao and Tseng 2004).

In the methods of mass customization, like design for variety (DFV), Quality Function Deployment (QFD) and House of Quality (HoQ) are mostly used. Analysis of the impact of external factors on the product structure is essential to define elements that are likely to change (Jensen et al. 2012).

3.3 Flexibility against design change

Flexibility is defined as "the ability to change and adapt a building to altered activities through its physical and administrative environment"(Greden 2005). Through achieving or increasing building flexibility, the life span of the building will be increased and the costs can be reduced without the need for an extensive work to make the adaptability of the building (Israelsson and Hansson 2009). The physical characteristic of a the building (product) describes the flexibility of the product that is sought against changes in use or requirements (Hansen and Olsson 2011). Three types of flexibility can be distinguished: Adaptability, convertibility, and expandability (Pati et al. 2008; Arge and Landstadt 2002; Bjørberg and Verweij 2009).

There are many factors that affect flexibility: Awareness aspects, finance aspects, future planning, installation, production, and material standard (Israelsson and Hansson 2009), as shown in Figure 3.6:

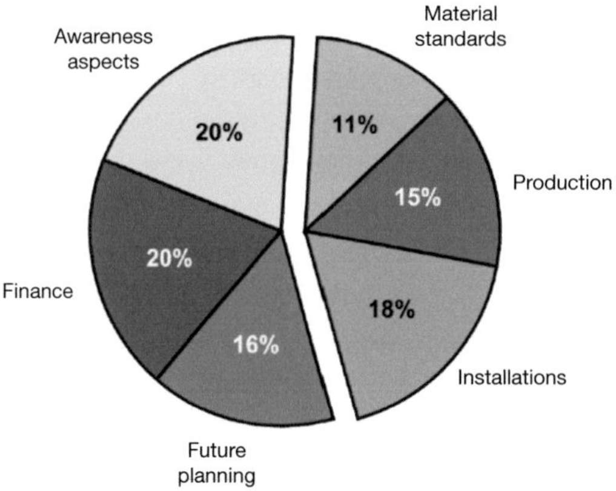

Figure 3.6: Factors affecting flexibility (Israelsson and Hansson 2009)

The various parties that could influence the decision making process to make flexibility (Israelsson and Hansson 2009) can be seen in Figure 3.7:

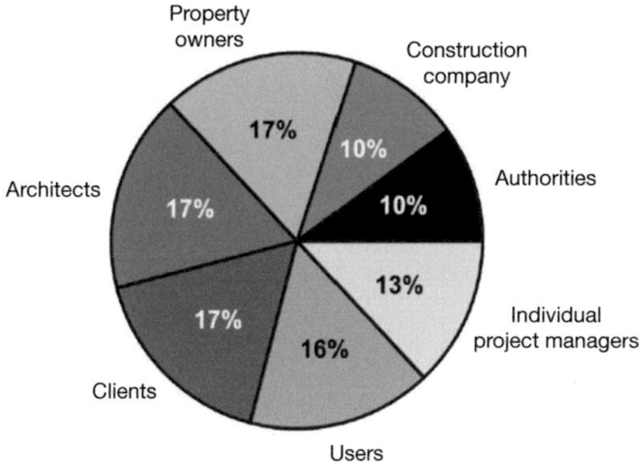

Figure 3.7: Decision makers affecting the flexibility (Israelsson and Hansson 2009)

According to Israelsson and Hansson (2009), factors that influence the flexibility can be classified into "hard aspects": Awareness aspects, finance aspects, future planning; and "soft aspects": Installation, production, and material standard. However, they do not explain what type of production and installations can support the flexibility, how they can be developed, and how they affect the development process during design. Israelsson and Hansson (2009) conclude that there is a lack of knowledge and lack of awareness regarding flexibility, and that improving quality of decision making assures that appropriate levels of flexibility will be provided.

Although increased flexibility of a building increases economical value through the reduction of reconstruction flexibility (Israelsson and Hansson 2009), the initial costs of production will increase less than 2 per cent, which can be recovered at the first renovation (Greden 2005).

4 Modularization and standardization

Modular product architecture can be viewed as a subset of product architecture, which is often approached with component standardization; however, modularization and standardization are not the same thing (Börjesson 2012). Figure 4.1 explains some main differences between the two concepts:

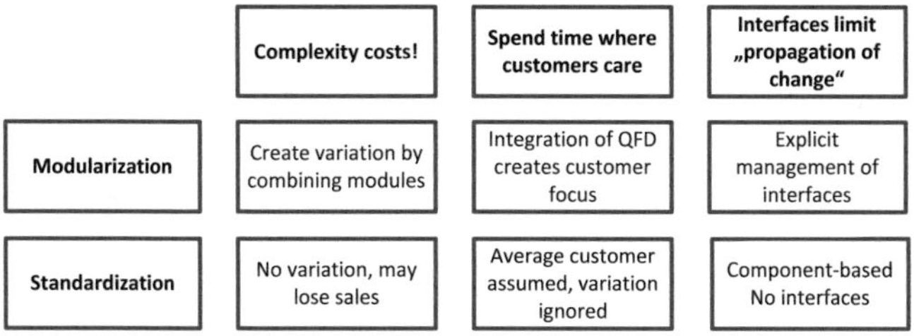

Figure 4.1: Modularization and standardization are not the same thing (Börjesson 2012)

In the next sections, concepts of M&S and their methods will be discussed.

4.1 Modularization

Modularization is a method to reduce the complexity of products by decomposing them into portions that can be managed efficiently (Baldwin and Clark 2003). Although the term modularization is often used in the literature, there is no consensus on the definition of this concept and the proper use of it (Gershenson et al. 2004a). Therefore, understanding the meaning of modularization is a core point when dealing with the utilization and benefits of the implementation. It can be stated that definitions and methods of modularity depend on its purpose (Hölttä-Otto 2005). Baldwin and Clark (2000) define modularity as "a design structure, in which parameters and tasks are interdependent within the modules and independent across them". The term chunk (module) is used for a major physical element of a product that could be shared among products to exhibit high levels of commonality (Jensen et al. 2012). According to Erixon (1998), "a module is a physical building block with standardized interfaces selected for company-specific reasons". The roots of modularization have been analyzed in previous research (Mohamad et al. 2013).

To achieve modularization, there are many methods and purposes which show the referred non-consensus in defining the concept "modularization" that deals with different contexts of dependency and similarity (Gershenson et al. 2004b). Gershenson et al. (2004a) emphasize the importance of product representation as the first step to achieve M&S. Different types of modularization will be explained in the next sections.

4.1.1 Types of Modularization

Three types of modularity can be distinguished according to their goal: Modularity-in-design, modularity-in-production, and modularity-in-use (Baldwin and Clark 2006). Baldwin and Clark (2006) claim that the goal of modularity affects the way the modules are structured.

4.1.1.1 Modularity in design (MID)

A product or process is modularized when "the elements of design are split up and assigned to modules according to formal architecture or plan" (Baldwin and Clark 2002; Baldwin and Clark 2006). Baldwin and Clark (2006) define modularity in design as follows: "modular-in-design is if (and only if) the process of designing it can be split up and distributed across separate modules".

The main purposes of modularization are (Baldwin and Clark 2006):

- To make complexity manageable;
- To enable parallel work; and
- To accommodate future uncertainty.

Modularization of a system includes identifying three elements of the system (Baldwin and Clark 2006): "(1) the architecture of the system: what are its modules, (2) the interfaces between the modules: how do the modules interact, (3) tests: how well do the modules perform their tasks, and how well do the modules work together".

To develop a modular structure, the first step is to analyze the dependencies between design parameters by using methods such as Design Structure Matrix (DSM) to represent the dependencies in the systems. In the next step, the task is to define modules within the system where the design parameters are more interdependent. Then, the dependencies between the modules can be modified to "design rules" that should be obeyed from the parameters of the independent modules. The separated units "or modules" must still be integrated into a functioning, whole system (Baldwin and Clark 2006). According to Baldwin and Clark (2006), a modular design structure has three characteristic parts: "(1) *design rules*, which are known and obeyed by teams responsible for individual modules; (2) so-called *hidden modules* that "look to" the design rules, but are independent of one another as work is proceeding; (3) and a *systems integration and testing module* in which the hidden modules are assembled into a system, and any remaining, minor problems of incompatibility are resolved". The modular system can also be represented using design hierarchy, as can be seen in Figure 4.2:

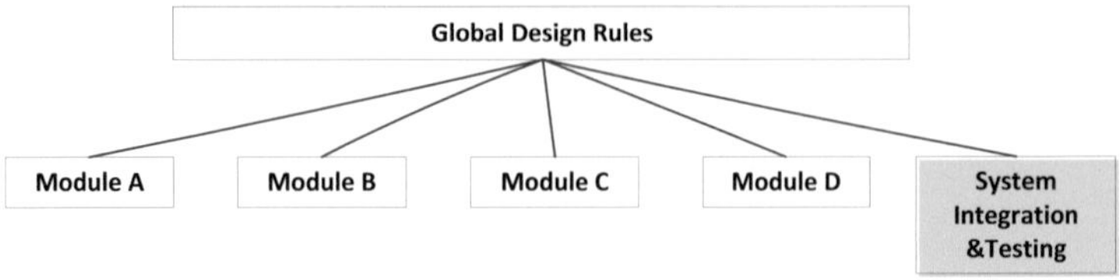

Figure 4.2: A Two-level Modular Design Hierarchy (Baldwin and Clark 2006)

4.1.1.2 Modularity in production (MIP)

Modularity in production (or process modularity): To achieve modularity in production, the specification of the components, for example, its dimension and functionalities, are design rules for the manufacturing process (Baldwin and Clark 2006). Modularity in production supports mass customization and can be characterized as process modularity (Sako and Murray 1999). Baldwin and Clark (2006) argue that modularity in production of a system does not mean that the design of the system is modular. In modularity in production, making products will be easier by dividing manufacturing process into process modules or cells (Baldwin and Clark 1997). Process modules could be a large production cell or a work station in an assembly (Sako and Murray 1999). Gershenson and Prasad (1997) define manufacturing modularity as "the development of product modules with minimal dependencies upon other components in the product with regard to the manufacturing process". They introduce a methodology for modular product design that depends on three issues: Attribute independence, process independence, and process similarity.

Lai and Gershenson (2008) introduced a type of modularity in production that depends on similarity and dependency for assembly modularity, and they defined assembly cost factors that include tool changes and fixture changes. Assembly modularity can be achieved through process based design, and it depends on the representation of dependency and similarity that impacts the assembly process (Lai and Gershenson 2008).

4.1.1.3 Modularity in use (MIU)

Modularity in use is defined as follows: "A system of goods is modular in use if consumers can mix and match elements to come up with a final product that suits their taste and needs" (Baldwin & Clark 2006). Modularity in use aims to decompose the product into components that can be made by different manufacturers but, at the same time, they fit together because they have standard sizes. Therefore, MIU supports customization (Baldwin and Clark 2006) and mass customization (Pandremenos et al. 2009).

The benefits that could be obtained from the different types of modularity depend on the type of modularity and the number of modules of the type of modularity (Sako and Murray 1999), as can be seen in Figure 4.3:

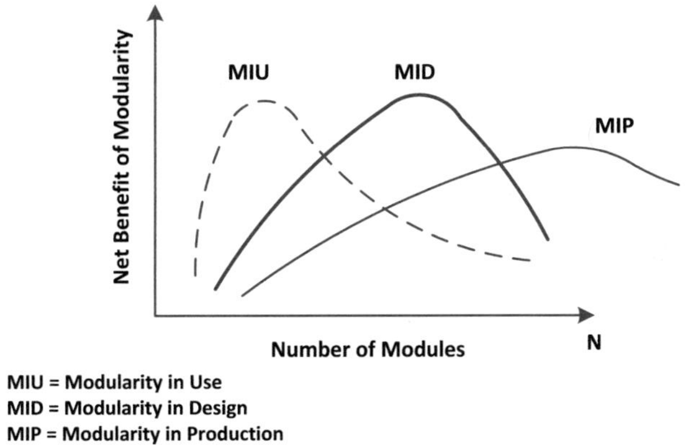

Figure 4.3: The effect of number of modules to the net benefit of modularity (Sako and Murray 1999).

Figure 4.3 shows that by modularity in production (MIP), higher number of modules will be needed to achieve high benefits as compared to MID and MIU.

4.2 Standardization

Component standardization is the use of the same component in multiple products and is closely linked to product variety (Ulrich 1995). Standardization can arise only when: (1) a component implements commonly useful functions; and (2) the interface is identical to the component across more than one different product (Ulrich 1995).

Ulrich (1995) compares the modular and integral architecture in their ability to enable standardization of components and he explains how the modular architecture enables standardization. Standardization of components has implication for the manufacturing firm in the areas of cost, product performance, product development, economics of scale, and learning (Ulrich 1995). While standardization of components may attract some firms, this standardization can cause more costs when the standard components give excess capacity that is not necessary. In this case, the firm may choose to adopt the standardization, but may be with justification because of economic savings resulted from reduced complexity, for example, in purchasing, quality control, inventory management and field services (Ulrich 1995).

4.2.1 Methods of standardization

According to Swaminathan (2001), there are four methods of standardization to make mass customization reduce the negative effects of increased product variety and variability: Part standardization, process standardization, product standardization, and procurement standardization.

4.2.1.1 Part standardization

Part standardization is using commonality in components or subsystems along the product line. The benefits that can be achieved are (Swaminathan 2001): (1) reduced cost because of economic of scale,

(2) reduced inventories because of risk pooling, (3) reduced part proliferation, and (4) improved predictability of requirements for components. Although there are many benefits of part standardization, the challenge of reducing customer value could appear through reducing product differentiation. Therefore, the trade-off between improving operation and reducing differentiation should be considered.

4.2.1.2 Process standardization

Process standardization enables postponement of the customization to as late in the process as possible, and it requires that the process be modular so the firm can store inventory in semi-finished forms (Swaminathan 2001). Some methods of process standardization are, for example, optimization of process sequencing, or designing platforms (as in the automobile industry) to enable producing different types of a product without major changes in the production line. The degree of process standardization depends greatly on product type and on where customization occurs in the product (Swaminathan 2001). Process standardization in design aims to postpone the decisions until the last responsible moment because of uncertainty about customer requirements, and to enable flexibility against possible changes in design.

4.2.1.3 Product standardization

Product standardization is the ability to offer high variety of products using a small inventory. In this case, the customer gets a product with a superset of features when he asked for a version of the product that is not available; therefore, this type of standardization could lead to dissatisfaction of the customer (Swaminathan 2001).

4.2.1.4 Procurement standardization

Procurement standardization can be achieved when a variety of products can be produced by common equipment or/and using common parts. The benefits of procurement standardization are improvement of resource utilization and reducing inventory through risk pooling (Swaminathan 2001).

It can be stated that the type of standardization depends on the modularity of the product and the process, as can be seen in Figure 4.4:

Modular	**Part Standardization** Maximize component commonality across products	**Process Standardization** Delay customiztaion as long as possible
Non-Modular	**Product Standardization** Carry a limited number of products in inventory	**Procurement Standardisation** Leverage equipment and part commonality across products
	Non-Modular	Modular

Process

Figure 4.4: Strategy of standardization (Swaminathan 2001)

Standardization includes implication for product design, process change (design process, production process, and procurement process), ability to meet customer value, and degree of outsourcing, whereas the ability of the firms to adopt standardization depends on the modularity of product and process, what it is trying to achieve for the customer, and the cost associated with standardization (Swaminathan 2001).

4.3 Approaches to modularization

An approach to modularity includes "the method by which the architecture is derived and the method itself is the way the data is captured and processed" (Börjesson 2012). Börjesson (2012) found that a cross-functional team is a very important success factor, as is "a solid management commitment". According to Hölttä-Otto et al. (2005) there are three main approaches to modularity: (1) Heuristics, (2) Design Structure Matrix (DSM), and (3) Modular Function Deployment (MFD).

4.3.1 Heuristics

According to Börjesson (2012), "heuristics try to capture how designers actually think". This method depends on separate modules from a single product's function structure by finding the dominant flow, branching flows, or conversion-transmission function pairs (Stone et. al. 2000), as seen in Figure 4.5:

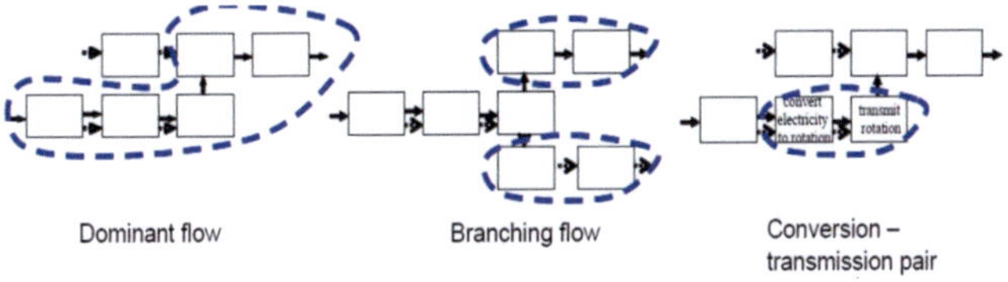

Figure 4.5: Function structure heuristics (Hölttä-Otto 2005)

To apply the function structure heuristics method, the function structure should be defined. The function structure diagram can be resulted differently, where "there is no single correct way of creating a function diagram and no single correct functional decomposition of a product" (Ulrich and Eppinger 2008). In the next step, the heuristics define possible modules, where many possible alternative modules can be defined by grouping functions according to the heuristics (Hölttä-Otto 2005). The heuristics are maximal heuristics which state only that one should not define modules larger than indicated; however, the defined modules, according to dominant flow, for example, can be subdivided (Hölttä-Otto 2005). The main modularization criteria considered in the function structure heuristic method are functionality and module interfaces (Hölttä-Otto 2005).

4.3.2 Modular Function Deployment (MFD)

Modular function deployment (MFD) is based on functional decomposition, such as functions structure heuristic method, but in this method, modularity drivers other than functionality are considered (Ericsson and Erixon 1999). According to Erixon (1998), MFD is "based on the idea of decomposing the customer requirements (CRs) into specific statements and linking them to measurable and controllable product properties, decomposing customer requirements of the product into technical solutions, describing how each technical solution (TSs) impacts the performance on a particular product property, and grouping technical solutions carrying similar properties and strategic intent to define modules"(Börjesson 2012), as can be seen in Figure 4.6:

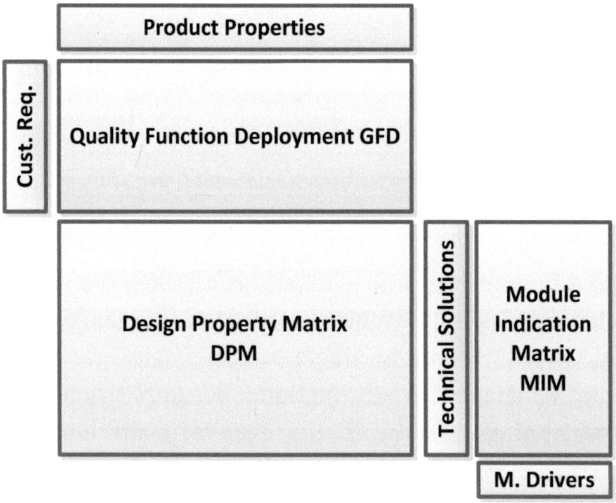

Figure 4.6: MFD uses three interlinked matrices (Börjesson 2012)

In this method, the grouping into modules can begin by defining the functions that have the highest summed scores, and the functions dominated by the same modularity drivers are good candidates for a module (Hölttä-Otto 2005).

4.3.3 Design structure matrix (DSM)

DSM can be seen as a method to map interdependencies (Börjesson 2012), and it can also be used to define modules within a single product's architecture (Hölttä-Otto 2005). Figure 4.7 shows an example of a component-based DSM. The task of the development can be sequential, parallel, or coupled according to the dependencies between the components represented by the mark x, as can be seen in Figure 4.7. Component-based DSM can be used to define modules in a product architecture (Hölttä-Otto 2005); for example, the component E and F are interdependent and can be defined as module.

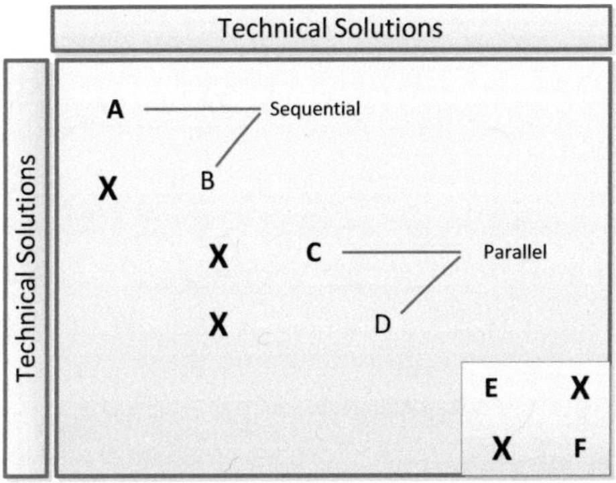

Figure 4.7: DSM is based on mapping dependencies (Börjesson 2012)

On the other side, DSM is used to manage the tasks of a development process or the teams by minimizing unnecessary design iterations during the development process (Ulrich and Eppinger 2004).

Defining clusters (or modules) in the DSM can be made by applying clustering algorithm to group components or functions in such a way that the interactions are maximized within the clusters and minimized between the clusters. The main task of the clustering algorithm is to reorder the rows and columns of the DSM such that the marks x are as close to the diagonal as possible. "The algorithm can result in overlapping modules or it may leave a function out of the final clustering, in which case it is up to the designer to decide how to handle them, for example, the overlapping section could be duplicated and placed in both modules or forced to be only in one of the modules where the algorithm suggested it could be" (Hölttä-Otto 2005). DSM can only be seen as a supporting and learning tool during the modularization process (Schmidt III et al. 2008). Schmidt et al. (Eppinger and Browning 2012), used DSM to modularize the building structure for adaptability purposes.

4.4 M&S in the construction industry

In the construction industry, M&S are used to decrease costs and improve quality by developing platforms that can be used over several projects (for example, Jensen et al. 2009; Jensen et al. 2012), i.e. industrialization of buildings. In Jensen et al. (2009), M&S are used to develop more flexible building systems over several projects using quality function deployment (QFD), where the focus is on modularizing and standardizing certain elements, like walls, and the feedback to improve the building system is obtained from project to project. Their model consists of the development of a technical platform and a configuration phase of a project. The technical platform can be described as the core product description system and it consists of standardized modules (elements). In the configuration phase, developed standardized modules are configured to give a customized product (like a wall) with alignment between four views: Customer view, engineering view, production view, and site view. The idea depends on separating the development of the technical platform from the configuration process to achieve customization. Figure 4.8 represents this developed model:

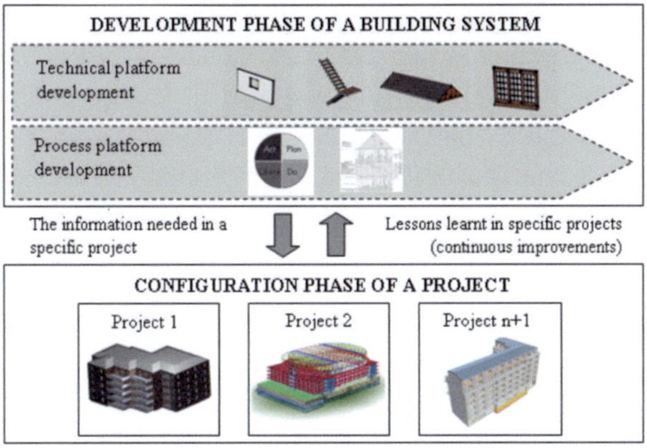

Figure 4.8: Industrialized housing process (adapted from (Lessing 2006)) (Jensen et al. 2009))

Veenstra et al. (2006) developed a methodology depending on methods of design for variety, and applied it for a single family housing. The methodology aims to develop a platform according to the concepts of open building.

Court et al. (2009a) used "modular assembly" to improve health, safety, and productivity in the mechanical and electrical construction. They defined modular assembly as "the ability to pre-combine a large number of components into modules and for these modules to be assembled off-line and then bought onto the main assembly line and incorporated through a small and simple series of tasks", where modularization can be achieved with or without off-site manufacturing capability (Court et al. 2009b).

Generally, M&S concepts are used in the construction industry to reduce design efforts while offering a variety of products in the follow-up products (product family) (Jensen et al. 2012). Also, modular construction is used mostly to describe the prefabrication of a volumetric or three dimensional parts of the building offsite, like mechanical systems, entire rooms, bathrooms, or kitchens to obtain savings in cost and time; however, the prefabrication of these parts has challenges and design requirements like structural and MEP design factors that must be carefully considered (Velamati 2012). This type of modularity has its benefits and challenges according to many factors, such as type of the building, transportation, and work strategies.The next sections describe the benefits of M&S in the construction industry.

4.4.1 Improving the design process: Flexibility

During the design process, a product may need to be redesigned because of design iterations and development of new versions of the product during design (Hölttä-Otto 2005). Modularity of product architecture is suggested to improve flexibility during design; this will enable the designer to tolerate high levels of risk (Thomke 1997). Upton (1994, 1995) defines flexibility as "the ability to change or react with little penalty in time, cost or performance" and he distinguishes between two types of flexibility: (1) the potential ability to produce a range of products (process range) and (2) the ability to quickly change between products (process mobility). By modularizing the product architecture, flexibility can be achieved through the robustness of a module (if changes should be made or because of design upgrade), where the interfaces (boundaries of the modules) should be designed properly (Hölttä-Otto 2005).

4.4.2 Improving the construction process

The linkages between product architecture and aspects of manufacturing were discussed in the literature extensively. The decisions in the earlier design and detailed phase of design have a great influence on the construction process (Emmitt et al. 2004). Takt time planning is one of the benefits of M&S, and according to interviews with construction firms that use takt time planning, the phase of the detailed design greatly impacts takt time planning of the construction process of MEP systems.

On the construction site, the sources of variation in cycle times for building spaces can be classified as production system problems, inherent work content variability, and external factors (Brodetskaia et al. 2012). The main sources of production system variability are (Koskela 1992): Insufficient materials, overcrowding of work areas, lack of information, inappropriate equipment, inconsistent deliveries, and low levels of control on availability of subcontracting teams. Each building space has particular finishing requirements and different volumes of each work type (Brodetskaia et al. 2012). According to Sacks and Goldin (2007), changing client requirements are the most significant external sources of variation.

Improving the construction processes of MEP systems through M&S of MEP systems is made through supporting takt time planning. Takt time planning depends on location breakdown structures, and the benefits can be gained through savings in time and costs, as explained in Figure 4.9 (Linnik et al. 2013):

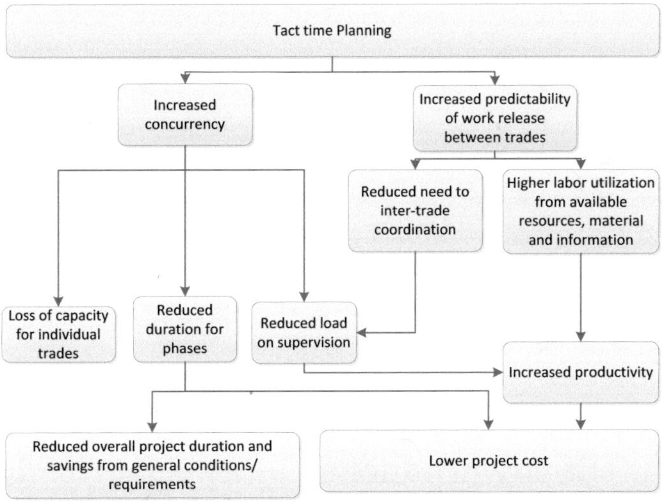

Figure 4.9: Expected benefits & costs from takt time planning (Linnik et al. 2013)

Uniqueness of material increases systems' complexity; therefore, reducing the variety of components improves the production system performance (Tommelein 2006). Tommelein (2006) distinguishes between "specific" and "standard" material to explain how standard material creates flexibility in a production system by mitigating a matching problem which is known in the AEC industry, where "availability of standard materials creates the opportunity to use any one in any one of several locations in the facility being built". She recommends the manager to design their project-based production systems by exploiting product standardization opportunities. According to Tommelein (2006), there is a relationship between the number of standard products used in each area of a facility and the time needed to complete the project; however, her model does not discuss the design requirements and conditions for that.

Demand variability, which is caused by changes in installation timing and sequence, and changes in design (Ballard and Arbulu 2004) will be reduced through standardizing components and structures of MEP systems.

According to a study made by Digitales Bauen (2008), M&S increase the required work to install the modules, but they reduce the waiting and search time and interruptions. Therefore, the reliability of the plan will be increased greatly. Increasing prefabrications' possibilities through M&S (because of savings resulted through prefabrication of standard components) contributes to reduce the interruptions and waiting time and searching for material, which are caused by storing the elements on the construction site and by prefabrication on the construction site to adjust larger elements.

4.4.3 Off-site production and prefabrication

M&S facilitate prefabrication and mass production. Off-site production (OSP) has been promoted as a way to improve performance of construction projects (Blismas et al. 2005). Egan (1998) emphasizes the important role of supply chain partnering, standardization and off-site production in improving the construction process.

Blismas et al. (2005) investigated the constraints of the implementation of OSP, and they defined four constraints in off-site production and the relationship between them: Value, process, supply chains, and knowledge constraint. Value constraints occur through an obligation, set by clients, to accept lowest cost options rather than best value, where process constraints occur through the client's or designer's inability to freeze the design and specification early enough within the construction project process to ensure that delivery of the component is made when required on site.

Table 4.1 explains the drivers and constraints of off-site production. It can be seen that some drivers and constraints relate to off-site production of large parts (for example, modules of the building) or to off-site production of any other parts regardless of their volume.

Table 4.1: List of drivers and constraints of off-site production (Blismas et al. 2005)

	Drivers		**Constraints**
	Cost drivers		**Site Constraints**
D1	Ensuring project cost certainty	C1	Restricted site layout or space
D2	Minimizing non construction costs	C2	Multi trade interfaces in restricted work areas
D3	Minimizing construction costs	C3	Limited or very expensive available skilled on-site labour
D4	Minimizing overall life cycle costs	C4	A problem transporting manufactured products to site
	Time Drivers	C5	Live working environment limits site operation
D5	Ensuring project completion date is certain	C6	Limitation to movement of OSP units around site
D6	Minimizing on-site duration	C7	Site restricted by external parties
D7	Minimizing overall project time		**Process Constraints**

	Quality Drivers	C8	Short overall project time scales
D8	Achieving high quality	C9	Unable to freeze design early enough to suite OSP
D9	Achieving predictability of quality	C10	Limited capacity of suppliers
D10	Achieving performance predictability throughout the lifecycle of the facility	C11	Not possible for follow-on projects to use the same processes
	Health and Safety Driver	C12	No opportunity for component repeatability on this or future projects
D11	Reducing health and safety risks		**Procurement Constraints**
	Sustainability Drivers	C13	Project team members have no previous experience of OSP
D12	Reducing environmental impact during construction	C14	Obliged to work with a particular supply chain
D13	Maximizing environmental performance throughout the lifecycle	C15	Not willing to commit to a single point supplier
D14	Implementing respect for people principles	C16	Obliged to accept lowest cost rather than best value
		C17	Key decisions already made preclude OSP approach
		C18	Limited expertise in off-site inspection
		C19	Early construction/manufacturing expertise and advice unavailable
		C20	Obliged to accept element costing based on SMM

A developed program to increase prefabrication in the Swedish government showed great impact on productivity and quality (Bertelsen 2005). A further advantage of modularity, also referred to by Ulrich (1995), is that modularity of the product allows the variety to be created at the final assembly (the last stage of production process) and then improves logistics and supply chains. According to interviews, variety of components of MEP systems reduces the opportunity of prefabrication of piping, where on-site work is preferred to deal with the uncertainties in systems' design and to reduce the costs of manufacturing a variety of components off-site.

5 Lean thinking

Lean thinking begins with understanding the value of a project and which activities and resources are necessary to achieve that value (Poppendieck and Poppendieck 2003). Hansen and Olsson (2011) underline the importance of applying lean thinking in building design to achieve the usability of the completed building that supports the core business, and that systems management is important to achieve the best outcome. Workflow is one of the core concepts of lean thinking, and it is described as "the progressive achievement of tasks along the value stream so that a product proceeds from design to launch, order to delivery, and raw materials into the hands of the customer with no stoppages, scrap or backflows." (Womack 1996).

Although the importance of managing a design process effectively and efficiently is to reduce uncertainty and improve quality, many efforts are still spen on the construction process to deal with challenges that should be dealt with in design (El. Reifi and Emmitt 2013).

5.1 Design management and lean design

The design process is characterized by "the translation and transformation of ideas, expectations, and user needs to find and evaluate multiple solutions to a problem" (Peña 1987). Hansen and Olsson (2011) describe the design process as an evolving process used to develop alternatives or design solutions that increase the understanding required in the following processes and tasks in design, and they define it as an iterative process, a cycle of discovery, where finding alternatives and evaluating them adds value (Hansen and Olsson 2011). The early phase is both a problem seeking- and a problem solving process (Parshall et al. 1987). Jensen et al. (2012) describe the design process as "an iterative sequence involving several levels where decisions on one level involve analyses of possibilities and consequences in other levels". Shamsuzzoha and Helo (2012) investigate the importance of information flow in defining the product architecture during design, and how the information exchange influences the basic architecture of the product development process. Ballard (2002) explains that "Analyzing the nature of the design process reveals that "rationalistic" models of problem solving processes are inappropriate for the design process, which rather oscillates between criteria and alternatives, as in a good conversation from which everyone learns".

One of the important features and challenges during design is that not all required information is always available and that decisions are not made in order to complete the design (Hansen and Olsson 2011).

Iterations in design is essential to find innovative and adequate concepts and designs (Hansen and Olsson 2011). In contrast to making, iterations in design are important to improve the customer value (Hansen and Olsson 2011). However, it must be distinguished between positive and negative iterations in design (Ballard 2000b). Positive iterations are required to increase understanding and make interpretations of the purposes, whereas negative iterations are unneeded iterations and are an important source of waste in design (Ballard 2000b).

Making decisions should be made at a suitable point of time during design to increase flexibility during design (Ballard and Howell 2003). However, realizing the design alternatives must be made through defined "lead time", where the decisions must be made within it. Freezing the design should be made at the last responsible moment. The last responsible moment is the "point at which failing to make the decision eliminates an alternative" (Ballard 2000b).

During the design process, many conflicts could emerge because of high number of participants and interests. It is normal to find conflicts and different perspectives on a project; for example, the user or owner are interested in organizational issues, while the general contractor is interested in technical issues (Blyth and Worthington 2010). The most important aspect of the decision making process is the conflicting interests between supply and demand; while demand aims to keep design options open as long as possible or necessary, supply tries to restrict the process as early as possible (Hansen and Olsson 2011).

Uncertainty in decision making is another important aspect in projects, and it relates to the gap between information needed and information already processed (Galbraith 1973). What decisions should be made and at what time during design is a very important aspect in managing the design process, where decisions that cannot be made or do not need to be made early should be delayed (Blyth and Worthington 2010). Blyth and Worthington (2010) highlight the importance of layered decision making strategy to define what decisions must be made and by whom.

Lean design refers to approaches, principles, and methods to manage design processes or product development (Jørgensen and Emmitt 2009). It considers the effects of product design on customer value and on the downstream life cycle process (Dombrowski et al. 2014). Lean design aims to reduce uncertainty in the design process and increase efficiency through increasing customer value (Hansen and Olsson 2011).

5.1.1 Flexibility in managing the design process

Uncertainty can be defined as "the gap between the amount of information needed to perform a task and the amount of information already possessed by the organization" (Galbraith 1973). Flexibility in product and process aims to reduce the effects of uncertainty because it enables changes and adjustments that emerged from uncertainty (Hansen and Olsson 2011). According to Olsson (2008), two types of flexibility in project management can be defined: External flexibility deals with what requirements will be met, whereas internal flexibility deals with how the requirements will be met. The similarity of the two levels of flexibility can be seen in lean perspectives, where external values relate more to the high priority of the end customer value, and internal flexibility relates to the efficiency in the developing process (minimizing waste through the developing process) (Hansen and Olsson 2011). Flexible processes and flexible robust designs are needed through the development process to enable changes and adjustments (Mikkelsen et al. 2003). Hansen and Olsson (2011) discuss the flexibility achieved through layered design process related to the layered product concept.

To define the boundary of the research more, it should be distinguished between two types of design change: 1) changes during design, and 2) design changes after the fact (in construction and operation phases). Developing strategies to manage changes during design aims to improve the design process itself and value generation (for example, last responsible moment), whereas developing strategies to manage changes during the lifecycle of the building relates more to structure and interdependencies in the product architecture; for example, reducing interdependencies through modular design. In this research, the flexibility we will talk about is product and process flexibility achieved through modularization and standardization.

5.1.2 Customer value

According to the literature, quality, cost and delivery time are the three parameters that represent the value of the customer. The customer-perceived quality should be distinguished from the engineering-achieved quality (Kano 1984).

Utility theory has been used to evaluate the customer perception of quality (for example: Thurston 1991; Malen and Hancock 1995). Du et al. (2006) introduce utility functions to quantify the customer-perceived value in terms of the quality utility per unit cost and the ratio of marginal utility to marginal cost. Ballard (2008) identifies value as that which allows the understanding of customer purposes "what they want to accomplish", and he argues that it is obligatory to architectural and engineering designers to consider the global customer.

5.1.2.1 Value and value generation (workflow)

Value, as defined in Lean Thinking (Womack and Jones 2003), refers to materials, parts, or products - something materialistic which can be understood and specified (Koskela 2004).

Value is "the end goal and therefore the establishment of value parameters at the outset of a project are keys to the achievement of improved productivity and client/user satisfaction" (Emmitt et al. 2004). There are many views and perspectives about value in the literature (Emmitt et al. 2004).

According to Rossi et al. (2012), five principles are the basis of the lean logic: "specify value, identify the value stream, make the value flow, let the customer pull the process and pursue perfection".

For buildings, the basic value structure is based on six key areas of value: Beauty, functionality, durability, suitability, sustainability, and build ability (Emmitt et al. 2004), which are values of clients that could represent owner, user, or society during the life of a building (Bertelsen and Emmitt 2005). The construction team also have their values but they should focus their efforts on achieving the value of the client (Emmitt et al. 2005). Emmitt et al. (2005) divide value into external and internal value, where external value is the clients' value and the value that the project should end up with, whereas internal value is the value that is generated by and between the participants of the project delivery team (contractor, architects, designers, etc.). An essential aim of Lean Construction is to aid in the delivery of external value by managing the internal value generation process (Björnfot and Stehn 2007).

The word value has two characteristics (Christoffersen 2003):

- The perception of value is individual and personal and is therefore subjective: Agreement of an objective best value for a group will differ from the individual's perception of value.

- Values will change over time.

Emmitt et al. (2004) state that value should be viewed as an output of the collaborative work of design and construction teams and as central to productivity, where a comprehensive framework of work should be provided.

The challenge in value management is when it is considered as an additional discipline to process management in design and not as an integral component of design management, and sharing of values is a challenge for individual organizations and temporary project groups (Emmitt et al. 2004).

The match between design and construction is essential in the value perspective, and it can be achieved effectively through engaging all stakeholders to define and confirm the values of the project (Emmitt et al. 2004).

Value is used as an indicator for the performance of an organization (Gidey et al. 2014). Continuous improvement efforts aim to improve value. Adding value or improving value generation is the target of the Plan-Do-Check-Act (PDCA) cycle (Deming's cycle) through continuous quality improvement. Process improvement is one aim of the PDCA and can be described through capturing and enhancing customers' specifications regularly (Gidey et al. 2014). PDCA achieves improvements not only in problem identification, but also through rationale flow of value (Gidey et al. 2014).

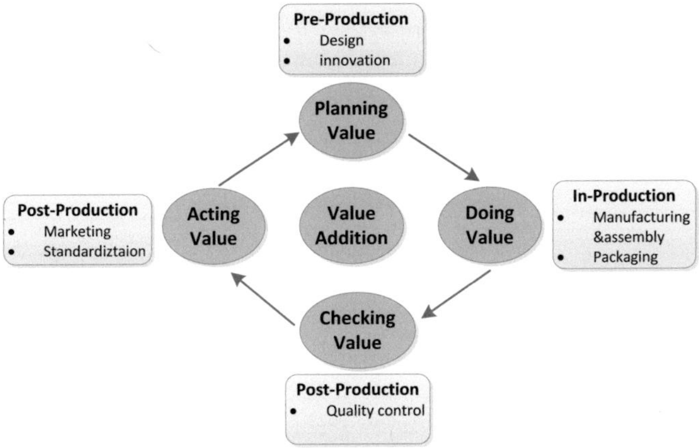

Figure 5.1: PDCA for value addition (Gidey et al. 2014)

Value addition can be achieved through the functions of the three phases, as can be seen in Figure 5.1 (Gidey et al. 2014):

- Pre-production: Innovation and design.
- In-production: Manufacturing, assembly, and packaging.
- Post-production: Quality control & inspection, standardization, and marketing.

There are many applications of the PDCA cycle: When starting a new improvement project; when developing a new or improved design of a process; when defining a repetitive work process; and when implementing any change (Zokaei and Simons 2006).

Quality function deployment (QFD) (Akao 1990) is a tool that is mostly used to capture customer value and map the customer requirements against product properties (Jensen et al. 2012). Emmitt (2006) found through his observations the importance of workshops in defining the parameters of value between all actors in the early phases of design.

From the lean perspective, managing the design process includes meeting the objectives of finding a design that satisfies the end customer/user, and of developing structures and systems that enable effective and stable workflow during the construction process (Hansen and Olsson 2011). Reducing waste during design process, like negative iterations is also an objective of managing the design process (Ballard 2000b).

The most used lean techniques to manage internal value or workflow control are (Björnfot and Stehn 2007): Last planner system, value stream mapping, just-in-time production and supply-chain management, and the Poka-Yoke or the five whys technique, target costing as an integrated internal/external value view. Set-based design (Ward et al. 1995) aims to keep the design solutions open as long as possible in order to delay the decisions to the last responsible moment.

Pull techniques are used to manage workflow in design as a production control strategy through pulling information during the engineering process, and it is used to match up the various elements needed to actually perform work (Ballard 1999). Regarding to the two primary techniques for the management of work flow, namely, push and pull (Hopp and Spearman, 1996), push systems assume infinite capacity, i.e., "should" disregards "can"; whereas in pull systems "can" overrides "should" (Ballard 1999).

The potential benefits of pull in design are "to manage the sequence and rate of production so as to provide maximal customer value while conforming to stakeholder needs and demands" (Ballard 1999); however according to Ballard (1999), the nature of the design process is a challenge to apply pull in design, where design alternatives and design criteria are interdependent (e.g., Austin et al. 1998).

Pull planning is a tool to define who is supposed to do what and when, and a tool to track commitments and to ensure that all prerequisites are identified (Tiwari and Sarathy 2012). Pull planning helps to discover misinterpretation of scopes of work between the team members (Tiwari and Sarathy 2012).

Pull planning creates an atmosphere of trust, transparency, and communication, which helps in achieving a comprehensive plan that facilitates identification of constraints and interdependencies (Tiwari and Sarathy 2012). Tiwari and Sarathy (2012) introduce an example of implementation of pull planning in the preconstruction phase as a strategy for collaboration between the design team and the construction team with the goal of producing design drawings to restrict any post-permit design changes due to cost, constructability, or coordination issues.

Application of pull techniques to design has two main challenges: The nature of the design process, and the traditional way in managing design where push is used (Ballard 1999). "Delaying decomposition of design activities until near in time to their scheduled execution is a response to the first obstacle. The second obstacle is to be overcome by demonstrating the effectiveness of combining pull with push" (Ballard 1999).

Pull is an integral part of the Last Planner system (Ballard 1999) which will be explained in the following section.

5.1.3 Last Planer System (LPS)

The last planner system (LPS) is a system to manage production, and it was used successfully to manage production in design and construction phases. Managing production process using LPS depends on defining the relationship between ends and means, where someone decides what work will be done tomorrow, called 'assignments' (Ballard 2000c). "The person or group that produces assignments is called the 'Last Planner'" (Ballard 2000c). The term Last Planner refers to "the hierarchical chain of planners, where the last planner acts at the interface to execution" (Koskela and Howell 2002).

The assignments are defined through communicating the requirements of the last planner, where possible differences between what will be done and what can be done and should be done may exist (Ballard 2000c, Koskela and Howell 2002), as can be seen in Figure 5.2:

Figure 5.2: The formation of assignments in the Last Planner planning process (Ballard 2000c)

Ballard (2000c) defines LPS as follows: "The last planner production control system is a philosophy, rules and procedures, and a set of tools that facilitate the implementation of those procedures". LPS can be described in terms of "principles that guide thinking and action, the functions it enables to be performed, and the methods or tools used to apply those principles and perform those functions" (Ballard et al. 2009), as follows:

1. Principles:
   - Plan in greater detail as you get closer to doing the work.
   - Produce plans collaboratively with those who will do the work.
   - Reveal and remove constraints on planned tasks as a team.
   - Make and secure reliable promises.
   - Learn from breakdowns.

2. Functions
   - Collaborative planning
   - Making Ready
     - Constraints identification and removal
     - Task breakdown
     - Operations design
   - Releasing
   - Committing
   - Learning
3. Methods and tools
   - Reverse phase scheduling (aka "pull planning", "pull scheduling", "phase scheduling"; stickies-on-a-wall)
   - Constraints analysis; constraint logs; risk registers
   - Task hierarchy: phase/process/operation/steps
   - First run studies
   - Daily huddles
   - Reliable promising
   - Metrics:
     - Percent plan complete
     - Tasks made ready
     - Tasks anticipated
   - 5 Whys analysis

The activity definition model is a technique in the look-ahead phase of the LPS to decompose the (design) activities to be performed according to their schedule. Figure 5.3 explains the activity definition model.

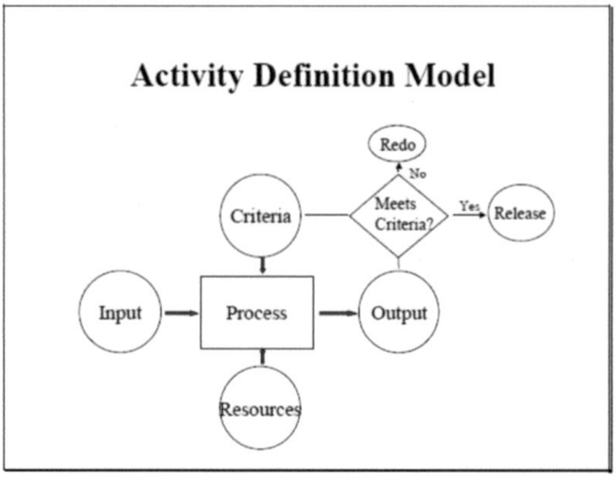

Figure 5.3: Activity definition model (Ballard 1999)

Design criteria are derived from customer requirements and should be used to produce the design. However, the design process should be seen as a value generating process through the progress of design (Ballard 2000c).

According to Ballard (2000c), there are two elements to explain the procedures of the LPS: Production unit control and work flow control. Some of the critical quality characteristics of an assignment on the production unit control are: (1) the assignment is well defined, (2) the right sequence of work is selected, (3) the right amount of work is selected, and (4) the work selected is practical or sound, i.e., can be done. Whereas production unit control relates to the work executed by the production units, workflow control relates to the work flow between the production units (Ballard 2000c). Figure 5.4 shows the five phases of LPS:

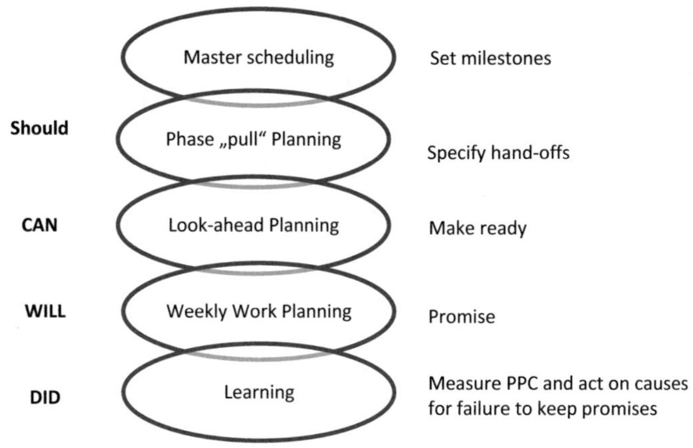

Figure 5.4: The Last Planner System of production control (Ballard 2000c)

The time horizon of the look-ahead planning is 3 to 4 weeks. In the look-ahead planning the assignment for upcoming assignments is made ready. This is a pull system (Ballard 2000c) which assures that the prerequisites for the assignments are available (Koskela and Howell 2002). "Should" represents the tasks in the plan, and "can" represents those tasks that could be realistically started in the situation (Koskela and Howell 2002).

In LPS, control consists of: (1) measurement of the realization rate of assignments (percent plan complete (PPC), and (2) investigation of causes for non-realization and elimination of those causes (Koskela and Howell 2002).

5.1.3.1 LPS in design

One of the challenges during the design process is managing workflow, where the assignment is very important in the management process (Ballard 2000a). LPS has been used successfully on the construction site to increase the reliability of workflows (Hamzeh et al. 2009); however, using LPS in design is challenging because of increased number of participants and increased complexity of coordination between the participants (Hickethier et al. 2013). Workflow during the design process comprises different types of variability, especially iterative processes between design alternatives and predefined owners' value (Ballard 2000; Ballard 2002; Hamzeh et al. 2009). As previously stated, during the production process in design conflicts may emerge between different customers or between producers; therefore, aligning the interests is important and trade-offs are unavoidable.

Hamzeh et al. (2009) show that LPS can be used in design to manage workflow, and they present the following practices of applying LPS in design according to Ballard (2000a) and Ballard et al. (2009): "(1) plan in greater detail as you get closer to performing the work, (2) develop the work plan with those who are going to perform the work, (3) identify and remove work constraints ahead of time as a team to make work ready and increase reliability of work plans (4) make reliable promises and drive work execution based on coordination and active negotiation with trade partners and project participants, and (5) learn from planning failures by finding the root causes and taking preventive actions".

Case studies of applying LPS to design show challenges caused by "failure to apply quality criteria to assignments and failure to learn from plan failures through analysis and action on reasons for plan failure" (Ballard 2002).

The conducted case studies and interviews in this research will show that using LPS during the implementation of M&S of MEP systems in design aims to improve the reliability of workflow, where a successful implementation requires discovering possible conflicts and removing them at the right time by the responsible persons. One goal of the management system to implement M&S is to reduce the conflicts without reducing the value or increasing the iterations, considering that M&S cause more iterations in comparison with a traditional design (without M&S).

M&S during design comprise many types of variability because they include more iterations. A pull system of the LPS can reduce the variability in the production process during design.

5.2 Integration of design and construction

Many of the problems that appear on the construction site are because of ineffectiveness in communication and decision making made in design, which is the result of uncertainty in production process (Emmitt et al. 2004) where production is seen as designing and making (Koskela 2000). Lean philosophy fosters integration of design and construction; however, in the lean literature there is a variety of using and understanding the integration of design and construction (Jørgensen and Emmitt 2009). Therefore, there is a need to define the lean approach of integrating design and construction more specifically (Jørgensen and Emmitt 2009). From the lean perspective of value optimization and waste elimination, four interrelated concepts of integrating design and construction can be explained (Jørgensen and Emmitt 2009):

- Aspects of vertical and/or horizontal integration in the construction supply chain and in between construction delivery and the management of real estate facilities and related services
- Integration of information systems for product and processes, which is often approached through a strong IT orientation
- Integration of working practices and collaborative processes in the construction project organization
- Constructability, which is often dealt with from the perspective of specific, practical advice for producing designs with a high level of constructability, e.g., the 'design for assembly' approach.

Faniran et al. (2001) introduced a model for a conceptual framework to integrate design and construction, as can be seen in Figure 5.5:

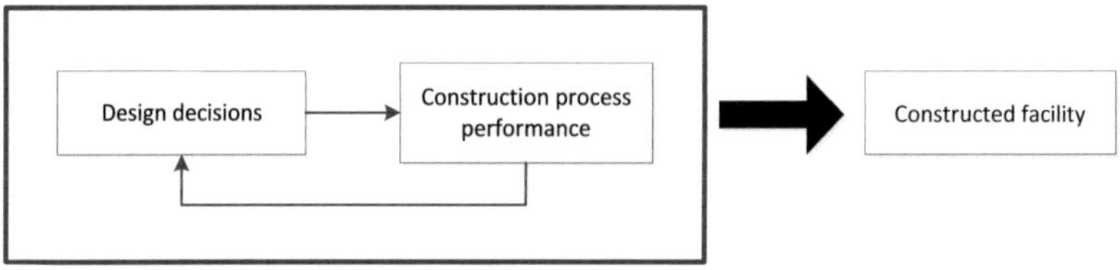

Figure 5.5: Conceptual framework of design/ construction integration (Faniran et al. 2001)

Integrating design and construction is a challenge from the perspective of ability to develop an intellectual argument and from the practical perspective to achieve real improvement (Emmitt et al. 2004). This challenge is about managing the interfaces (boundary conditions) between individuals and organizations through communication, cooperation, competences, and integration of customer values (Emmitt and Gorse 2003; Emmitt et al. 2004).

Jørgensen and Emmitt (2009) state that the integration of design and construction is highly impacted by contextual factors that cannot be ignored. Baiden et al. (2006) suggest that this integration is "the merging of different disciplines or organizations with different goals, needs and cultures into a cohesive and mutually supporting unit", and they describe the "integration construction project team" as "a highly effective and efficient collaborative team responsible for the design and construction of a project", where integration means "various skills and knowledge, and removes the traditional barriers between those with responsibility for design and construction in a way that improves the effective and efficient delivery of the project". Jørgensen and Emmitt (2009) define some issues that impact integration of design and construction: (1) Project value specification; (2) active client, user and stakeholder involvement; (3) decision and decision process transparency; (4) transparency regarding value/waste consequences of design decisions; (5) management of design iteration processes; (6) collaborative design with contractor/supplier involvement; (7) commitment from project participants (including suppliers); and (8) project team learning.

Malmgren et al. (2010) define four views of the product: Customer view, engineering view, production view, and assembly view, and suggest that the connection between the four views, i.e., information transfer, is an area of improvement. Figure 5.6 shows information flows between the different views according to Olofsson et al. (2010).

Another important challenge in the integration of design and construction is achieving the benefits of industrialization because of the need for an effective coordination between design, planning, and construction (Koskela 2003), where the clients, architects, structural engineers, contractors, sub-contractors, suppliers, and facility operators often have conflicting interests (Lu et al. 2011).

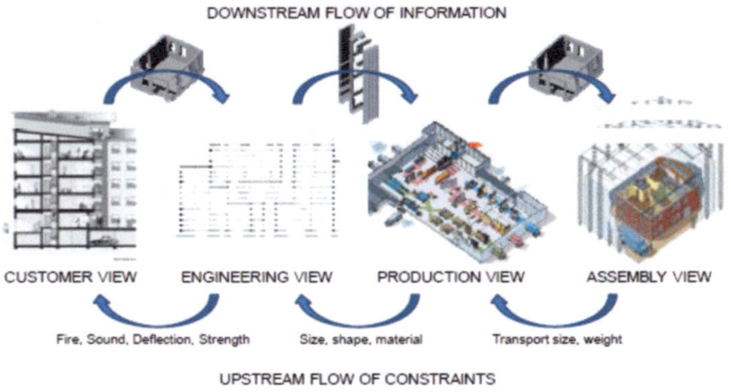

Figure 5.6: Information flow between the different views (Olofsson et al. 2010)

Lean philosophy is proposed to achieve integration of design and construction to align the interests of the project participants (Jørgensen and Emmitt 2009). This research explains the need to integrate design and construction while implementing M&S to achieve a successful implementation in design and construction.

5.3 Lean product development (LPD)

Product development "is the set of activities beginning with the perception of a market opportunity and ending in the production, sale, and delivery of a product"(Ulrich 1995). The focus in the lean product development process is to achieve a smooth flow of information and determine the root causes of non-value-added activities (Oppenheim 2011).

Ohno (1988) introduced seven major types of wastes during the manufacturing phase: Overproduction, waiting, transportation or conveyance, processing, inventory, defects and corrections, and motion. Womack and Johns (1996) introduced underutilized people's abilities as a new type of waste. Liker and Morgan (2006) reinterpreted the types of waste in the new product development as follows: (1) Overproduction, (2) waiting, (3) transportation or conveyance, (4) processing, (5) inventory, (6) defects and corrections, (7) motion, (8) unused employee creativity. These previous types of waste were used as references to explore types of waste recognized by the project teams in a case study conducted for this research.

According to Ward (2014), the main focus of lean product development is on value, i.e., generating a usable knowledge and profitable operational value streams which require to integrate the basic three kinds of learning: (1) integration learning (through integration of people), (2) innovation learning, and 3) feasibility learning to make better decisions. Existence of usable knowledge helps in making good decisions during development.

Ward (2014) defines three main causes for waste in product development: (1) scatter, (2) hand-offs, (3) wishful thinking, where hand-off is, like overproduction, the main source of waste that causes other types of waste during development, and it can be described as a separation in knowledge, responsibility, action, and feedback. Value in the product development process can be defined as "the right information products delivered at the right time, to downstream processes/customers, where it is quantified by form, fit, function and timeliness of information products" (Lai 1998). The value stream in the product development "consists of tasks that transform information and allow for the convergence of the segmented information to define a final design" (Walton 1999).

The Lean Aerospace Initiative (LAI) Product Development Team defined seven wastes in product development (Walton 1999):

Over Production • Too much detail • Unnecessary information. • Redundant development (Re-use not practiced)	**Waiting** • Information created too early • Late delivery of information • Unavailable information • Quality suspect
Transportation • Information/Software incompatibility • Communications failure • Not standards based • Multiple sources • Incompatible destinations requiring multiple transport **Defective Product** • Quality lacking or suspect • Conversion error • Wrong level of information • Incomplete information • Ambiguous information • Inaccurate information • Tolerance exceeded • Poor configuration management	**Processing** • Unnecessary serial processing • Lack of needed information • Poor/Bad decisions affecting future • Excess/custom processing • Not processed per process • Too many iterations/cycles • Unnecessary data conversions • Excessive verification • No transformation instructions • Decision criteria unclear • Working with wrong level of detail • Propagation of bad decisions • Processing of defective information • Multitasking when not required

Inventory	**Unnecessary Movement**
• Too much information • Incomplete content • Poor configuration management	• Information user not connected to sources requiring manual intervention • Information pushed to wrong people

Siyam et al. (2015) confirm the idea that research on product development (PD) focuses on value perspective during the product development process, and they state that value from product development can be described as:

"The degree to which a capability satisfies all relevant stakeholders, is delivered to them according to product or service quality, cost, and timeliness requirements, and is developed by performing effective and efficient processes that design and produce the satisfying capability within their budget and time constraints".

By defining the product development process (PDP) as a "channel connecting design process participants with other system stakeholders, or design activity as the channel connecting producers and recipients of value", the value perspective in the product development process can be described through three issues: Value definition, value creation and value delivery (Siyam et al. 2015).

5.4 Lean project delivery system

Ballard (2008) advances the hypothesis that "facilities better fit for purpose can be provided at less cost through rigorous project definition and through lean design and construction, i.e., through the lean project delivery system". Figure 5.7 represents the lean project delivery system (LPDS) developed by Ballard (2000 and 2006).

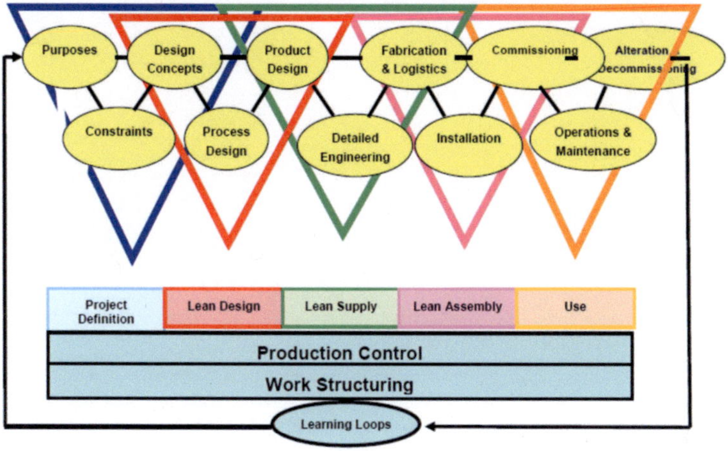

Figure 5.7: Lean Project Delivery System (Ballard, 2000 and 2006)

Ballard gives an example for such facilities, namely, healthcare facilities. In Ballard's hypothesis, the costs are "the relative costs of designing and constructing healthcare facilities pales in comparison to the costs of operations and maintenance", as can be seen in Figure 5.8. He gives further examples for such facilities when designing for sustainability, where shift of focus from first (capital) cost to whole life costs and outcomes is required, as appeared in publications like (Saxon 2005).

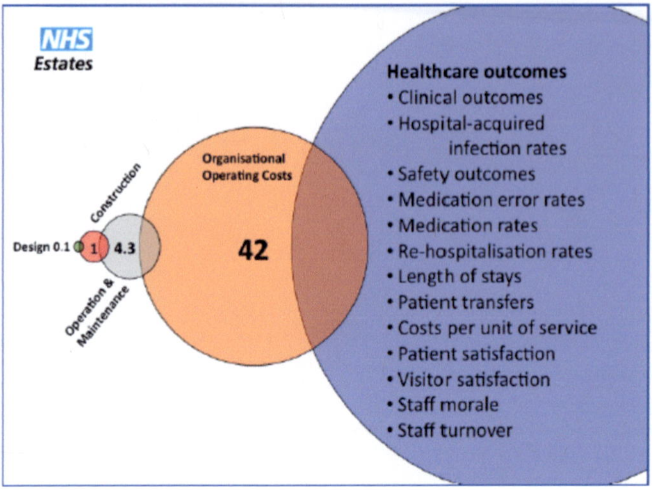

Figure 5.8: Relative costs (Ballard 2008)

Based on the recommendation of Matthiessen and Morris in their study published in 2004 (Langdon 2004), namely, that the first cost to the whole life cost should not be disregarded, Ballard (2008) suggests that elimination of waste can help designing and constructing better buildings for less costs through rigorous project definition, lean design and construction (the previous hypothesis). Rigorous project definition in the LPDS is made through a conversation between ends, means and constraints, where architecture, engineering, and construction (AEC) teams can help the customers to define what they want. Target setting is the first step in the design phase of LPDS, where target costing is used to deliver the customer value within constraints, and improvement cycles are essential to learn and improve performance. Figure 5.9 represents a developed model to apply target costing.

Ballard (2008) defines obstacles in the application of target costing, such as the inability for money to move across internal organizational boundaries between those responsible for capital costs and those responsible for business use of facility. An example for this obstacle appears during the application of M&S of MEP systems, where design costs increase because of increased, required efforts in design, while construction and operation costs decrease considerably. Moving money between design, construction, and operation departments could be used for defining target costs.

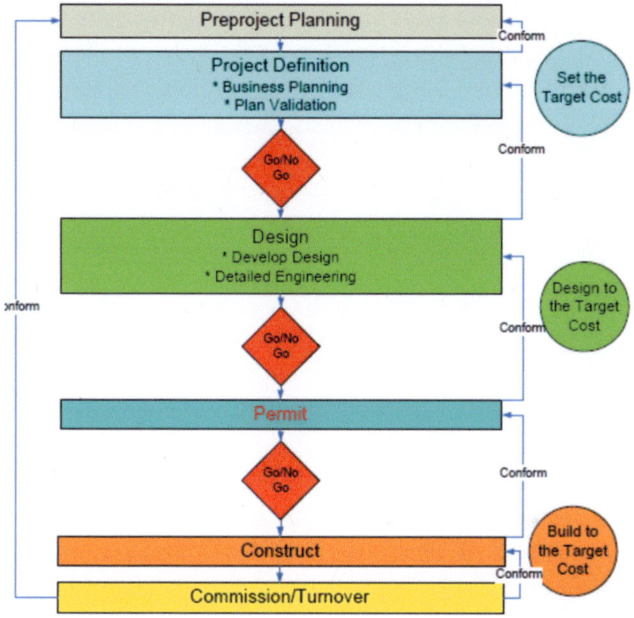

Figure 5.9: Project phases and target costing (Ballard 2008)

Maximizing value and minimizing waste at the project level are the goals of lean. However, these goals are difficult to achieve when the contractual structure "inhibits coordination, stifles cooperation and innovation, and rewards individual contractors for both reserving good ideas, and optimizing their performance at the expense of others"(Matthews and Howell 2005). The four major systemic problems with traditional contractual approach are presented by (Matthews and Howell 2005), and they show why the goals of lean are challenged, and why the project partners are prevented from organizing themselves to function as a single company with unified goals and objectives. The four problems are: (1) good ideas are held back, (2) contracting limits cooperation and innovation, (3) inability to coordinate, and (4) the pressure for local optimization. Traditional contracts fail to align incentives; therefore, they encourage local optimization (Ballard 2008). In contrast, IPD contracts do not have incentives to hold back ideas, but they have contractual incentives that reward cooperation (Matthews and Howell 2005). Integrated Project Delivery (IPD) is "a relational contracting approach that aligns project objectives with the interests of key participants, and it creates an organization able to apply the principles and practices of Lean Project Delivery System"(Matthews and Howell 2005).

5.5 Building Information Modeling (BIM)

BIM is seen as a tool in the lean construction community. However, many studies analyzed the supporting role of BIM in achieving lean construction, such as Sacks et al. (2010b) who introduced BIM as a platform to visualize workflow that enables pull flow and collaboration between teams in- and off site. Sacks et al. (2010a) analyzed the interactions between Lean Construction and BIM to conclude that these interactions can be exploited to improve the construction process beyond the degree to which it might be improved by application of either of these paradigms independently. Sacks et al. (2010a) present the functionality of BIM in the different project stages. Table 5.1 presents the functionalities of BIM in the design and fabrication detailing stages as it is defined in Sacks et al. (2010a).

Using of functionalities of BIM during the implementation of M&S will improve efficiency greatly. However, how the utilization of BIM impacts exactly the implementation of M&S from the lean perspective relates more to the research area of analyzing interactions between

Lean and BIM, and it requires the inspection of practical utilization of both BIM and M&S, which is not available in the current implementation of analyzed M&S. Therefore, this inspection will be outside the focus of this research, but it will be mentioned in the model to the necessity of applying BIM during the implementation of M&S.

Table 5.1: Functionality of BIM in the design and fabrication detailing stages (Sacks et al. 2010a)

Stage	Functionality
Design	**Visualization of form**
	Rapid generation of multiple design alternatives
	Re-use of model data for predictive analyses • Predictive analysis of performance • Automated cost estimation • Evaluation of conformance to program/client value
	Maintenance of information and design model integrity • Single information source • Automated clash checking
	Automated generation of drawings and documents
Design and Fabrication Detailing	**Collaboration in design and construction** • Multi-user editing of a single discipline model • Multi-user viewing of merged or separate multi-discipline models

After introducing the theoretical part related to M&S in product development, M&S concepts and methods, and lean perspective to manage design and product development process, the next chapter will introduce and discuss the conducted case studies in order to answer the research questions.

6 Case Studies

6.1 Case study 1: Modeling of design methodology of M&S of MEP systems

6.1.1 Background

The original idea of the analyzed design methodology is the developed model of Armilla, which returns to Professor Haller (1985). It depends on case-based reasoning and similarity concepts (FABEL-Report No. 131993). In FABEL-Report No. 13(1993), the developed method based upon Haller's model is defined as follows:

(1) How to define an adequate case representation (2) How to recognize and retrieve similar cases (3) How to find ways to adapt solutions. The method depends on describing asymmetrical geometry that allows adaption and then flexibility. The asymmetrical geometry helps to define asymmetrical layouts for the technical systems, which are very important in the adaptability of the building. The application of these structures is suitable for certain types of buildings, such as industrial buildings, schools and offices. Haller's system is based upon modularizing of the building service sub-systems such as water-sewage system, ventilation, air conditioning, heating and electrical cables as complementary to his industrialized min, mid, max structural system (Bock and Linner 2010). The advantages of designing Haller's system are as follows (Bock and Linner 2010):

1. Systemizing and modularizing the building's installation systems;
2. Supporting of industrialized pre-fabrication; and
3. Giving the overall building component system the potential of rearrangement and/or extension.

Other advantages are ascertained from interviews with the designer of a current project (case study 7):

- M&S of MEP allow making late changes in design with minimal design efforts. This is important because the owner is typically unable to define his wishes precisely in the early phases.
- They allow conversion of the building without the need for emptying the whole building; rather, it will only be necessary to empty part of it, and this will save huge costs during the operation of the building.
- It allows significantly reducing construction time, which means high benefits for the owner.
- By applying of M&S, planning time can be greatly reduced. More efforts will be needed at the beginning to align all customer values (owner, design factors of MEP, utilization), although design efforts will subsequently be greatly reduced.

Moreover, the benefits in improving the construction process are as follows:

- Pre-fabrication of standardized components and modules.
- Simple installation and management process on the construction site.
- Simple logistic and supply chains; for example, lifting equipment such as cranes will not necessarily be needed, and lifts could be easily used.

Furthermore, according to the same case study, some of disadvantages in design are defined as follows: more efforts in design, and over-design. Over-design is not only necessary in MEP systems but also in spaces where it may be needed to use larger spaces for utilization, which are not completely necessary ($1m^2$ or $2m^2$ of spaces are not really needed) to achieve standardization of MEP systems.

The analyzed M&S in this research is especially important when new technologies (high-tech or performance added sub-systems) have to be integrated in a built environment where the different economic and technological life span between basic structures of the building and the technical systems is extremely high (Bock and Linner 2010).

Further developments based upon Haller's model and how to apply them have been captured through discussions with the team of an engineering office that implements them. The following section describes the design methodology accordingly.

6.1.2 Design methodology

A methodology to modularize and standardize the MEP systems has been modeled. The model can be divided into two main phases: (1) Basics; and (2) M&S of MEP systems, as shown in Figure 6.1:

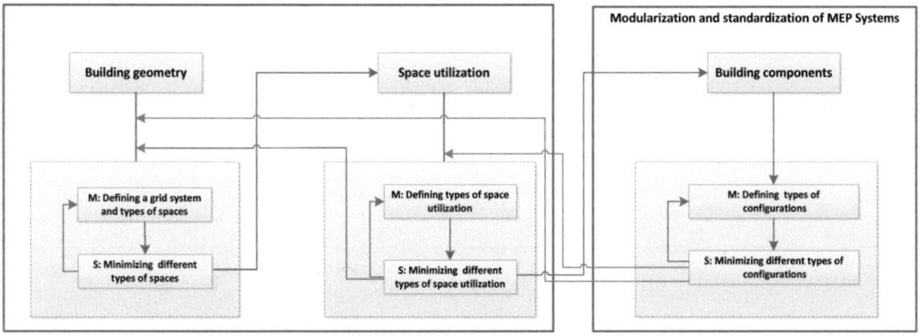

Figure 6.1: Design methodology for modularization and standardization of MEP systems

The basics phase addresses the reciprocal dependencies between MEP systems' design (design factors of MEP systems) and geometry and space utilizations (architecture of the facility). This phase represents the design phase.

The phase of M&S of MEP systems addresses M&S of MEP systems in terms of their structure and dependencies with other building systems and components. This phase represents the production system design.

Two types of workflows must be considered through the implementation: (1) the workflow within one phase of the model; and (2) the workflow between the two phases of the model.

In the following sections, we will explain the design methodology according to interviews with the designer and by means of an example:

6.1.2.1 Building geometry

Modularization started by defining a grid system for the building. The size of grid units is standardized and determined through the area in the geometry that allows for the maximum number of identical spaces in the building. Figure 6.2 shows above the grid system in green below the building. Positions of typical elements of the building geometry – such as columns, facade elements or shear walls – help during grid definition.

In the modularization process, the goal is either to completely put an element into one field of the grid – whereby the interfaces of the element align with the gridlines (e.g. facade elements) - or to put the element on the gridline, whereby the element becomes part of the interface (e.g. columns).

Standardization starts with grouping similar fields into 'types', e.g. the grid fields located in the corners of the building are similar because they have outside walls on two sides, thus constituting a type of field. Next, designers align the structure of fields of the same type by making small changes in the building design, e.g. moving a column onto the gridline between two fields. In order to minimize the number of types, it may be necessary to change the earlier defined grid system.

It can be noted that the value of M&S against the impact of changes on the design quality of the building must be weighed by the stakeholders (or customers).

6.1.2.2 Space utilization

The modularization process begins with assigning a category of space utilization to each field of the grid system. The goal in the modularization process is to align boundaries of spaces that have different utilizations with the interfaces between fields of the grid. An example can be seen in Figure 6.2 above, showing the different utilizations of building space in shades of grey.

During the standardization process, the goal is to minimize the number of categories and maximize the alignment between the types of grid fields and the categories, e.g. all corner fields of the building shall fall into the same category of space utilization. This process can include changes to the categories of space utilization, as well as changes to the utilization of spaces.

It can be noted that the standardization process may include risks of defining repeated units of utilization; however, the customers (stakeholders) must be involved to weigh the value of standardization against the impact of changes in the utilization of building spaces on design quality.

6.1.2.3 Systems' components

Modularization begins by assigning systems' components to fields of the grid. The goal during the modularization process is to align boundaries of systems with boundaries of fields of the grid. For example, changes in the diameter of ducts lay on the interface between two fields, whereby each field contains a minimum number of different types of duct. The above side of Figure 6.2 shows the structure of building components in the grid and the different types of spaces, including different configurations of components.

During the standardization process, the goal is to minimize the number of different configuration types for each type of field, as well as across different types of fields. For example, using a larger duct diameter than necessary in some parts of the building enabled a greater standardization of components. The customer must be involved to weigh the benefits of standardization - such as easier construction operations - against the higher costs for materials.

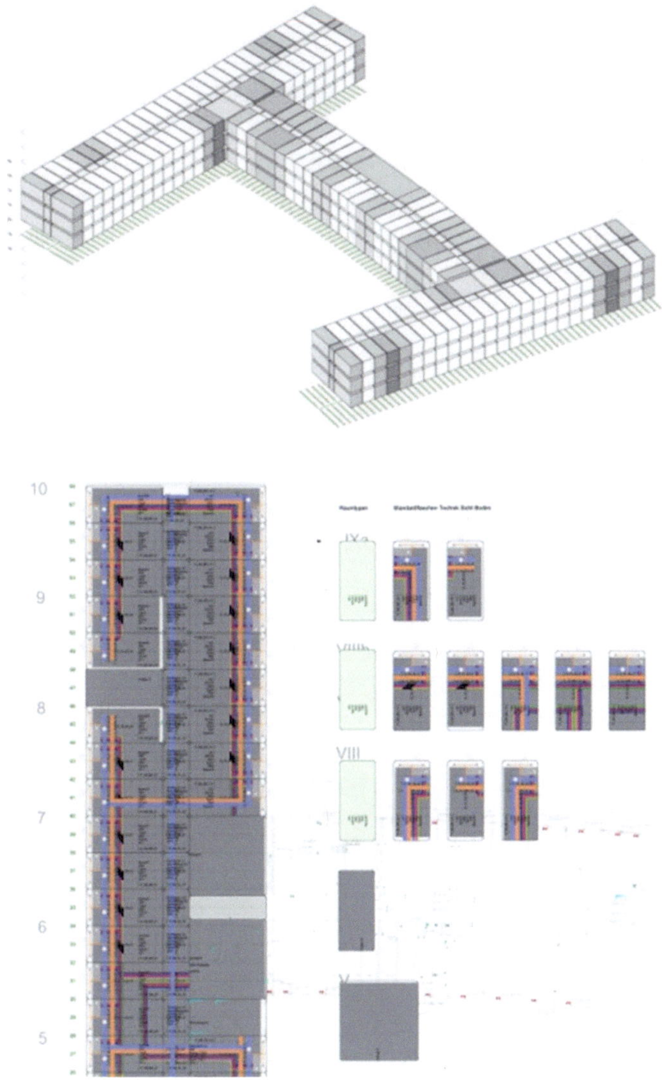

Figure 6.2: Structure of building geometry and space utilization (above); structure of building components (below)(source: Digitales Bauen, Karlsruhe)

6.1.3 Discussion

The modeled design methodology depends on a gradual application of M&S concepts during the design process: first on the geometrical model, then utilization aspects and subsequently detailed engineering design. This is close to the concepts of layered-implementation, which aims to reduce iterations through the implementation, and it requires understanding the interdependencies in building systems' design, tasks and organization.

The implementation of modularization is supported by modularization concepts in literature that depend on mapping functions to physical components (as explained in the utilization phase). Some procedures to improve modularization could be explained; for example, the repositioning of some elements like columns being included in one module (space) or exactly on the boundaries between two modules, i.e. the interface between the modules.

In the geometrical model, defining the size or dimension of the geometrical chunks (modules) may require developing criteria or tools to facilitate this process. In other industries, there are some tools and methods for this purpose, although the applicability in the construction industry should be investigated or adjusted to be applicable.

In the "utilization" phase, the categories of utilization should be applied to every field. Therefore, it is important to use a method to capture and organize this information. In any case, the user should be integrated in this process and he should participate in the negotiation to analyze the possible impacts on aspects of utilization. One or more alternatives could be developed in the standardization process; therefore, developing criteria (for example, investment perspectives) to evaluate the alternatives will be beneficial. The process of modularization by defining functions (utilizations) according to defined spatial chunks is aligned to modularization concepts in the literature.

The "developing of systems' components" phase represents the production system design. To standardize the components of MEP systems between the defined chunks, there may be one or more possibilities for this standardization. Therefore, using decision-making methods in this process will give the participants of this process the possibility to evaluate alternatives and make informed decisions depending on the possible alternatives of standardization; for example, using more material to standardize two or more chunks of MEP systems requires considering the "cost of used material" as a criterion in the decision-making process. Nevertheless, the question remains concerning who are the participants in this process, as well as whether they can be predefined.

During the modularization process, defining boundaries of MEP systems that corresponded to boundaries of geometrical and utilization modules is a challenge because: (1) different designers are responsible for designing different systems, and the different designers have different grid systems and software; and (2) restricting the adjustments only according to the boundaries of geometrical and utilization modules or defined boundaries of MEP's modules may require displacements or adjustments in other building elements.

During the standardization process, standardization is undertaken for the interfaces and the components. Using the same components for the modules requires the same spaces and utilizations, which requires finding small spaces and utilizations that are the same within the building or could be rendered the same through design (this prompts the need for more material and acceptance of the owner and other designers, who may need to make adjustments in their systems). Standardizing interfaces gives the ability to change parts or subsystems without changing larger parts, which may require adjustments and over-design.

The design methodology shows the importance of aligning customer values during the implementation. Affecting customer values can occur during the implementation of the design methodology and while executing the outputs of the design methodology - i.e. during the construction process - due to developing configurations and architectures that influence the construction and installation process. The alignment of values during the design and construction requires defining the customers and their values at the right time to assure the achievement in design and reduce iterations. This means that the alignment of values should include customers from the design and construction phases, as will be discussed in the next case studies. Indeed, the ability to align the customer values strongly depends on analyzing how they are affected during the design and construction, as well as finding methods to improve this alignment.

Table 6.1 shows aspects of the implementation and the potential of improvement of this case study depending on lean perspectives. The potential of improvement is defined depending on an analysis of this case study and requires further research. However, the focus of the research is placed upon developing guidelines to manage the implementation process.

Table 6.1: Aspects of implementation and potential of improvement (case study1)

Aspects of Implementation	Aspects of Improvement
What: Design Methodology of M&S	
Why: To analyze the adaptability of M&S and discover its implications.z	
Findings: As in other industries, modularization depends on mapping functions to physical components.	Inspecting the utilization of developed modularization methods in other industries such as a design structure matrix (DSM) and quality function deployment (QFD) which will facilitate the implementation.
Categories of utilization should be applied to every field.	User should be integrated in this process and should participate in the negotiation to analyze the possible impacts of M&S on aspects of utilization.
• M&S are iterative processes because they require considering their impacts on other buildings' systems and customer values. • Many alternatives could be developed during M&S processes.	• Using a set-based design will reduce negative iterations included in the iterative processes. • Using LPS will improve the production system control during implementation, through the reliable promising cycle and the commitment to implement.
There are interdependencies between different project partners during the implementation of M&S.	Transparency, controlling and commitment are required to improve performance during implementation (using LPS).

The design methodology shows the importance of early thinking about M&S during design to reduce iterations and re-work. The following case study explains the importance of beginning the implementation in the early phases of design. Before this, the delimitation of the analyzed design methodology will be briefly discussed in the next section.

6.1.4 Delimitation: Comparison between M&S in construction industry and studied research methodology of M&S.

The following table shows the delimitation between the analyzed method of M&S and other methods used in the construction industry as found in the literature:

Table 6.2: Delimitation between the analyzed method of M&S and other methods in the literature

Phase of implementation/ criteria		**Other methods of M&S**	**Analyzed method**
Construction	**Concept**	• Industrialized building • Applied on the whole building or geometrical parts of it	M&S are applied on MEP systems
	Benefits	• Mass production • Pre-fabrication	• Tact planning of MEP • Pre-fabrication of MEP • Mass production of MEP
Design	**Benefits**	• Mass customization • Flexibility against changes due to new technologies and reaching the complete operation time of building components	• Flexibility against changes in utilization of spaces • Customization • Flexibility versus changes during the design of the building (e.g. layout)
	Methodology	Developing a platform for a product family	• Applied to one project • Adjustments of aspects of: geometry, utilization and engineering components

In the construction industry, M&S concepts are used - as previously explained - to improve the construction process in terms of mass production and mass customization, as well as improving the design process by improving the potential of design for variety. Mass production and mass customization are used on the level of parts of the facility like rooms or walls, whereby the main process involves partitioning building elements to be produced off-site. M&S was also used to create variety while reducing design efforts in the development process of the family products. The focus of these methods is placed upon the facility's parts, where changes caused from the user perspective can be defined and integrated through QFD methods and tools.

Applying M&S methods to systems like MEP systems was not found in existing publications, aside from the work of Peter Court (2009) in the terms of modular assembly. However, design conditions, challenges and requirements were not analyzed in his work. M&S of MEP systems will offer the potential to improve the production process on the construction site in terms of mass production, as well as improving the ability to enable takt time planning for these systems and offering the potential to improve flexibility versus design change. The value of the product can be increased based upon the properties of the developed modularized and standardized structure, like the ability of splitting, substitution and others with little effort. In addition, M&S of MEP systems improve productivity on the construction site through reducing the variety of components.

6.2 Case study 2: Early implementation and optimization of the grid system

This case study shows the importance of beginning the implementation of the modeled design methodology (in case study 1) during the early stages of design. Defining the dimensions of the building and optimizing the dimensional system in the early stages shall be coherent with the process of M&S. This coherence ensures that modules, interfaces and components of the building can be clearly defined, which is essential in the M&S processes of the MEP systems.

Figure 6.3 shows one floor of the building after the process of M&S had been implemented. In the early phase, the form of the building and geometry were only sketched. Based upon the sketch, by implementing the above-described methodology, the dimensions of the building were adjusted by -12.5 cm and +5cm, as shown in Figure 6.3. This adjustment led to a symmetrical layout of the building, which in turn improved the potential for the standardization of dimensions of the layout within the same floor of the building. Standardized spatial areas then facilitate the definition of standardized modules for MEP systems across the floor, which also led to a reduction in the variety of MEP systems' components in one floor.

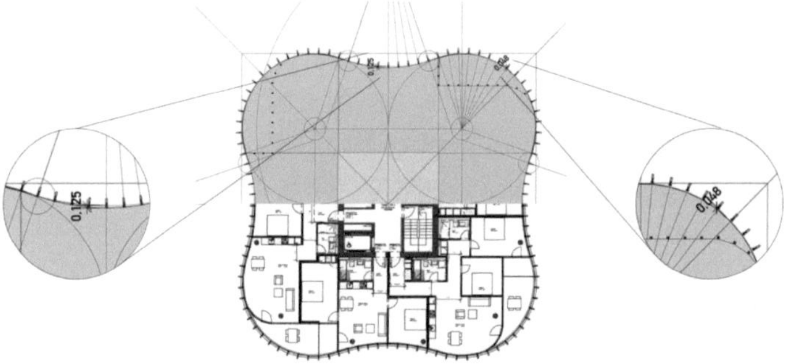

Figure 6.3: Adjustment of the building dimensions as a result of the design methodology (Source: Digitales Bauen. Allmend Wohntürme Luzern CH 2011. Marques Architekten Luzern CH. Halter Generalunternehmung Zürich CH)

6.2.1 Discussion

This case study shows the importance of early thinking about M&S of MEP systems to avoid possible iterations caused by late thinking. Responsible persons for the precise description of geometry must be available during this phase of the project. In addition, any conflicts caused by this process in the early phases should be considered. Unfortunately, details about the conflicts that might occur could not be captured. The main challenge is the early thinking about M&S, which requires an understanding and willingness to implement and eliminate conflicts. To achieve this, the specification of the value of M&S should be made explicit at an early stage to all stakeholder to be supported by the participants.

The customers in this process are the persons who will be affected during this process. Questions that should be answered are: (1) Who will be affected during this process? (2) How can the affected persons be made available during this phase? (3) How can the value of these customers be aligned while adjusting the dimensional system? All the affected persons should be integrated in this process to define the impacts and acceptable adjustments. Therefore, the impacts of this process cannot be defined in advance because they depends on the stockholder, including their priorities and values. Figure 6.4: explains this case study:

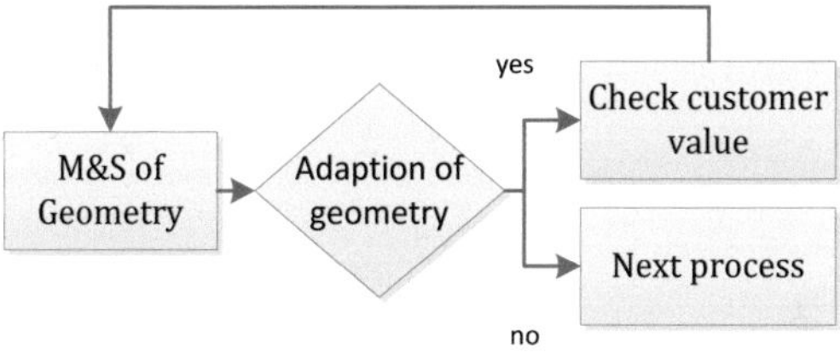

Figure 6.4: Early implementation and optimization of geometry

The results of this case study are aligned with the literature, suggesting that the ability of a design team to make an optimal design conceptualization early in the process is essential to the whole benefit cost analysis (Manavazhi and Xunzhi 2001), as well as reducing negative iterations in design (Ballard 2000b).

It can be concluded that the early implementation is a challenge because it requires defining M&S as value that should be supported by the participants. Moreover, this early implementation reduces iterations later during the implementation. Table 6.3 shows aspects of implementation and potential of improvement of this case study depending on lean perspectives.

Table 6.3: Aspects of implementation and potential of improvement (case study 2)

Aspects of Implementation	Aspects of Improvement
What: Early implementation of M&S.	
Why: Early implementation in design to improve downstream tasks is an aspect of lean.	
Findings: Early implementation reduces later iterations and leads to early adjustment of the building geometry as a prerequisite (or preparation) for M&S of MEP systems.	
Early implementation is a challenge because it requires defining M&S as a value that should be supported by the participants.	Integrated project delivery is required to ensure transparency and a willingness to cooperate as well as analyzing the impacts of M&S early on the business plan.

The next case study shows the potential benefits of the design methodology in reducing the variety of components of MEP systems.

6.3 Case study 3: Reducing the variety of MEPs' components

The goal of this case study is to discover potential benefits of M&S of MEP systems in reducing the variety of the components, which is one of the causes of instability of workflow on the construction site (Tommelein 2006).

In this case study, the results of applying the previous design methodology on an existing building design are presented. A comparison of the original and modular designs shows the potential benefits of the methodology. The subject of the comparison is a sector of the building of 1000 m^2 (Digitales Bauen 2008). It must be noted that this comparison is theoretical, because the modular design was not built. Nevertheless, the comparison shows that the methodology reduced the variety of MEP systems' components and structures.

6.3.1 Ventilation system:

The comparison between the original (before) and new (after) designs is based upon three criteria: (1) the number of leaps in the height of the ductwork (before 21, after 0); (2) the number of changes in the ducts' dimension (before 27, after 14); and (3) the number of different components of the ventilation system (before 48, after 14).

6.3.2 Connection components of ventilation, exhaust and fire protection systems:

The comparison criteria were: (1) the number of different outlets (before 43, after 36); (2) the number of outlets located on the interface between two modules (before 15, after 36); (3) the number of outlets not located on the interface between two modules (before 28, after 0); (4) the use cases of standardized structures (before 11, after 33); (5) the use cases of special structures (before 33, after 0); (6) the number of different structure variants (before 25, after 3); (7) the number of cases with easy installation conditions for outlets (before 26, after 36); and (8) the number of cases with difficult installation conditions for outlets (before 17, after 0).

In this comparison, the reduction of component variety is assumed to have little effect on end-customer value, while the changed 'look' of the MEP systems is hidden in the suspended ceiling. Table 6.4 shows the comparison between the results of original and modular designs:

Table 6.4: Comparison of outputs between original and modular design (Mohamad et al 2013)

MEP element	Criteria	Before (original design)	After (modular design)
Ventilation canal system	number of leaps in height of ductwork	21	0
	number of changes in ducts' dimension	27	14
	number of different components of the ventilation canal system	48	14
Connection components of ventilation, exhaust and fire protection systems:	number of different outlets	43	36
	number of outlets located on the interface between two modules	15	36
	number of outlets not located on the interface between two modules	28	0
	use cases of stand-ardized structures	11	33
	use cases of special structures	33	0
	number of different structure variants	25	3
	number of cases with easy installa-tion conditions for outlets	26	36
	number of cases with difficult installation condi-tions for outlets	17	0

The modular design has not been realized, and it was made only to show the potential benefits in improving the construction process by reducing the variety of the components.

6.3.3 Discussion:

In the previous study, the defined criteria explain the benefits of using M&S in terms of reducing the variety of components, although there is no further information about challenges or other factors such as costs. Some of the previous criteria used to compare original and modular design will be discussed.

The position of the outlets on the interfaces enables standardizing the distances between them, whereby the distances will be the same. This may not always be possible because this depends on the architectural requirements and conditions, and adjusting these requirements and conditions may affect other customer values (e.g. user/owner, dependent design systems) and should be aligned with them. This case study - where placing the outlets on the interfaces did not influence the customer value in terms of design quality - cannot be generalized, and it could cause conflicts between the participants during implementation.

The variety of the components like "number of leaps in height of ductwork" are caused due to dependencies with the structural and architectural models. Other criteria such as "number of changes in ducts' dimension" are caused due to restricting the capacities of the main ducts in the original design, whereby the "reduced number of changes in ducts' dimension" in the modular design results due to standardizing the dimension of the parts of the duct in the different utilization spaces, and then increasing the capacity of ducts. These two criteria again explain the modularization of the system through reducing the dependencies between MEP systems and other building systems through over-design, for example. Over-design requires more material. Reducing dependencies between MEP systems and other building systems to standardize these systems may require certain procedures for the design solution of the other building systems, which possibly prompts conflicts between the participants in design.

Typically, the designer is unable to evaluate some criteria that relate to the installations; for example, the criterion "good installation conditions". The installer, who will install the components, will be able to define the good installation conditions. The benefits in improving the construction process that could be achieved from the modularized and standardized structure of MEP systems strongly relates to the possibility of realization, which in turn relates to the construction teams. However, the need to integrate the participants of the construction phase contrasts with the current bid system, which prevents communications between the designer and construction teams.

The following chart in Figure 6.5 shows the variety of the components of the original and modular designs for the first three MEP elements studied in this case. Improving the construction process in terms of improving the learning process, reducing the installation time, improving the pre-fabrication possibilities and the impacts on material costs requires further empirical analysis, which exceeds the scope of this research.

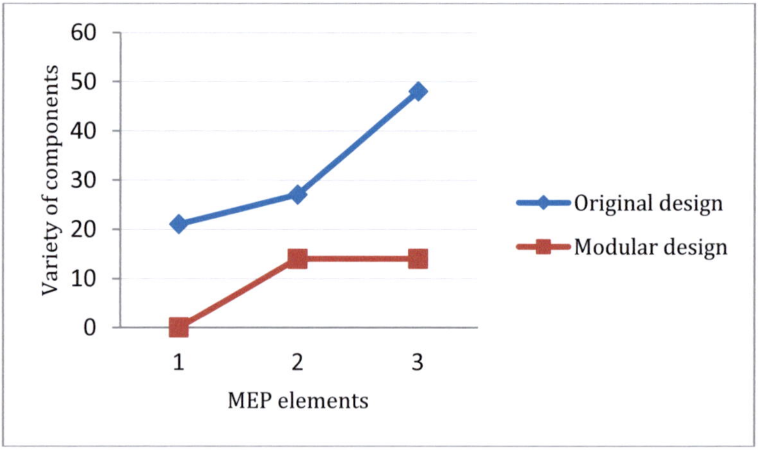

Figure 6.5: Variety of the components of original and modular design

We conclude that the design team's assumption that the modularized and standardized MEP structures has little effect on customer value cannot be assured, because there are dependencies between geometrical and utilization aspects from one side and M&S of MEP systems from the other. Moreover, there are dependencies between M&S of MEP systems in design and the conditions of realization on the construction site. All these dependencies can affect the possibility of implementation of M&S during design and construction. Therefore, these dependencies should be analyzed and managed during design to align all customer values, as seen in Figure 6.6.

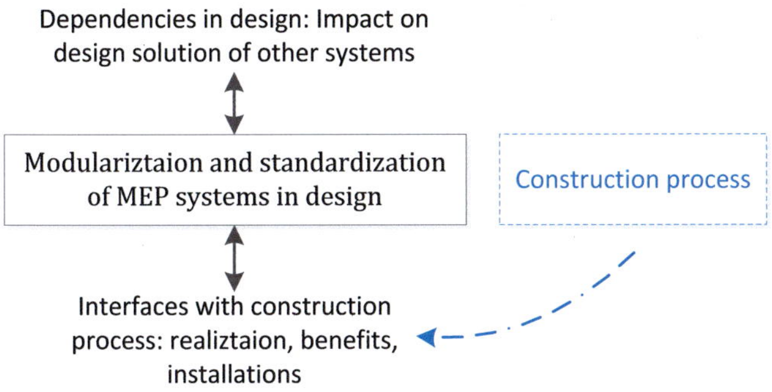

Figure 6.6: Consideration of design and construction dependencies during M&S of MEP systems

Table 6.5 shows aspects of the implementation and potential of improvement of this case study depending on lean perspectives.

Table 6.5: Aspects of implementation and potential of improvement (case study 3)

Aspects of Implementation	Aspects of Improvement
What: Benefits of using M&S in reducing the variety of components of MEP systems (theoretical).	
Why: Showing benefits of M&S is essential as a motivation to implement them. However, these benefits could not be realized due to refusing the design methodology and its requirements (e.g. cost overruns, conflicts).	Using TVD as a driver during implementation as an opportunity to reduce costs.
Findings: Assumption of the design team that the modularized and standardized structures of MEP have little effect on other customer values, cannot be assured.	The alignment of customer values should be assured to achieve end-customer value (cost, quality).
There are dependencies between M&S of MEP systems in design and the conditions of realization on the construction site.	Integrating of construction teams during implementation is required to evaluate the criteria of the decision-making related to construction process transparently.

6.4 Case study 4: Challenges during developing of installation supports

The case study shows the need for cooperation between the designer and the installer to achieve M&S of MEP systems. In addition, the case study shows some challenges during the phases of design and construction while applying M&S. In this case study, we have interviewed the facility manager, the plumbing and heating designer and installer, who has developed installation supports to enable standardization during design and construction.

The specialty of this case study is that the installer of the heating and plumbing was integrated during design due to a private relationship with the owner of the facility. The installer was involved during the design, albeit without any payments. The participation of other installers or construction firms during the design was not possible due to the system of competitions used (the bid system).

6.4.1 Type of the facility

The facility is an IT service office building. In this type of facility, there is normally a strong need to adapt the offices to fit the new needs of IT services. Adapting the offices includes changing the number of users in the offices or changing the dimension of the offices or their arrangements due to new technologies and work conditions.

6.4.2 Need for M&S

The previous design methodology was used for M&S of MEP systems. During the design process, one task was defined as determining where the pipes of the MEP systems must be placed to de-install them and reinstall new pips that fit in the old places when changing the design of the facility. To achieve this task, dividing the piping structure into small identical parts facilitates changing the design by replacing only the parts of the piping that should be changed when the design changes. The pipes must be precisely designed to be installable exactly as they are designed. The precise design and installation is also important for pre-fabrication of the piping and the ability to identify their location exactly in the future. M&S is achieved through splitting the structure of MEP systems to reduce dependency between parts of the structure to change some parts of the systems without changing the entire system. The standardization of parts of the piping structure reduced the additional costs (due to mass production and pre-fabrication) that could be incurred due to the need for more material (due to over-design). Pre-fabrication was important to assure delivering precise standardized parts, as well as improving productivity on the construction site, i.e. pre-fabrication was important from the perspective of the facility manager and the installer to weigh with possible additional costs caused by the need for additional material, as well as accelerating the installation process of MEP, thus reducing the costs of the construction process.

6.4.3 Development of the installation supports

As previously mentioned, the installer (who was also the sub-contractor) of heating and plumbing was integrated in the design due to his private relationship with the owner, although he did not receive any payment for this. During the design process, M&S were defined as goals that must be realized. One of the main challenges was that some designers were not satisfied with the goals of M&S, prompting a re-work of the design. According to an interview with the sub-contractor, we could learn more about one case of dissatisfaction: the designer was dissatisfied that the developed modularized and standardized structure of the pipes cannot be realized on the construction site, because - from the perspective of this designer - narrow available spaces will be available, which may prevent the installer from installing the pipes when placing them according to the modularized and standardized structure. Integrating the sub-contractor - who had long experience with installation processes - led to developing installation supports that enabled the later installation of the developed structure of the pipes precisely in the narrow spaces. The ability to install the pipes in narrow spaces was a requirement in some positions to achieve standardization of the pipes' diameters and lengths in design and construction, which facilitated off-site pre-fabrication.

In some positions and according to the modularized and standardized structure, the pipes must be placed in different angular situations. This gave cause to argue (by some designers) that realizing the structure will not be possible, although the installer could satisfy this by explaining how to bend the installation supports to realize the developed structure in these positions. This helped the designer to pursue the standardization of the pipes during design. Figure 6.7 shows the installation of pipes of different MEP systems in narrow spaces with the help of the developed installation supports.

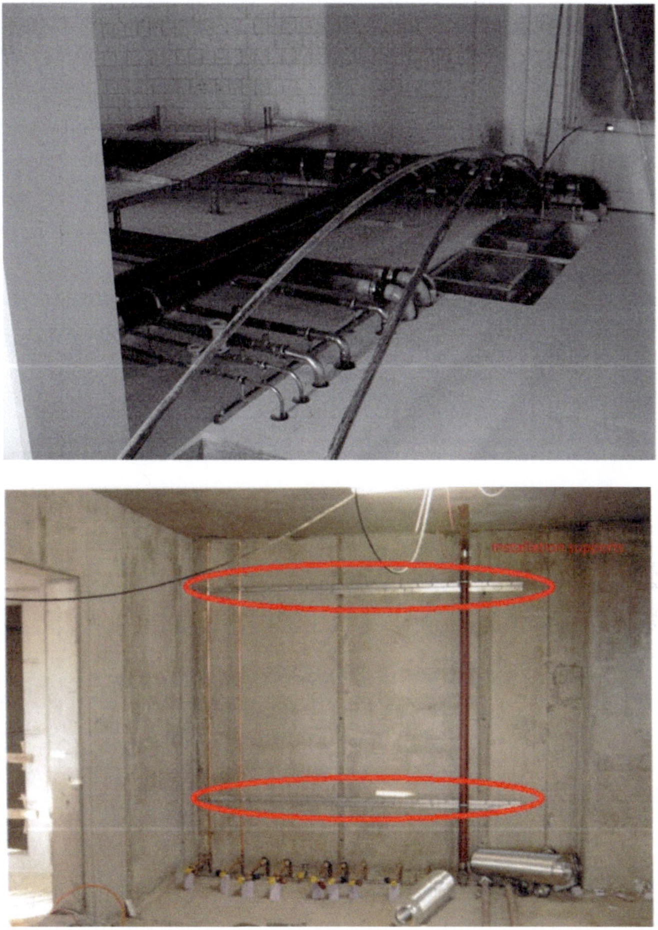

Figure 6.7: Installation of pipes of different MEP systems in narrow spaces with the help of the developed installation supports (above) and the installation supports (below)

6.4.4 Challenges

The challenges are defined through interviews with the facility manager and the plumbing and heating installer (also the sub-contractor). The following main challenges were encountered during the design and construction phases:

1. The coordination between the designer, and the sequence of design activities: the coordination between designers was important to avoid conflicts in design and on the construction site. This coordination was a challenge because there was a strong need to coordinate a high number of systems' interfaces; therefore, there were frequent iterations during the implementation. To standardize the structure of the MEP systems, any conflicts with other building systems must be avoided. During the M&S process - for example - sometimes the structural engineer should use more reinforcement to enable placement of some pipes in a certain place, or the MEP designers must place their pipes elsewhere due to restrictions related to the structural system.

2. Clear definition of user requirements: this was difficult on this project because the user was not integrated in the developing process, and the owner was unable to define them precisely in the early design. This caused iterations during implementation. In addition, this could reduce the chances of improving M&S, because there could be a need to change some aspects of utilization to improve M&S during design, which can be better evaluated by the user.

3. Conviction of M&S: there was no conviction of the idea of modularizing and standardizing the MEP systems. This caused conflicts, iterations and force (as expressed by the facility manager) during design. For example, there were difficulties with the electrical system designer, who argued that

if he placed the cable according to M&S structures, difficulties would occur in some positions due to the installation. However, there were no later problems on the construction site with the installations of the cables in these positions.

4. Having the right people during design: the participants must have a readiness to cooperate. According to the interview with the facility manager, "we have achieved this with force".

5. Changing the structure during construction phase: as an example, the electrical cables were installed differently by the installer and not according to the developed structure during design. The cause was that the installer did not have sufficient understanding and acceptance to the design and he thought he could make the installations better according to his experience. This caused re-work through de-installing the cables and re-installing them again according to design.

6.4.5 Discussion

The following points emerged from the case study:

- The dissatisfaction of the modularized and standardized structure appeared due to changing the way in which the designers should work. The designers had to cooperate with others and structure their systems in a different way. There was also no incentive to achieve these conditions.

- Integration of the knowledge of the installation was important for: (1) convincing the designer of the constructability of the system; and (2) developing installation supports to be delivered on time to the construction site and to facilitate construction process. Many types of the installation supports were developed. The relationship between the freezing of the design and the manufacturing of the installation supports lies beyond the scope of this research and should be analyzed in further research. This is aligned to empirical studies suggesting that the early integration of suppliers in product development can foster innovations in terms of configurations concerning how the components are linked together (Bozdogan et al. 1998).

- There were problems with the electrical system designer because the benefits of the system were unclear to him. In addition, his argument about possible difficulties of installation shows that he does not have sufficient knowledge about the installation process, and that knowledge of installation is necessary during implementation of M&S to avoid conflicts. Developing a way of making the goals of the project clear and creating satisfaction among the participants was required.

- One factor of successful implementation was that the installer/sub-contractor had his own factory to produce the required pre-fabricated standardized elements. Therefore, there was no need to coordinate manufacturing and delivering the pre-fabricated elements with additional teams. According to an interview with MEPs' chief in another design firm, this factor could be the cause for refusing some construction firms to make the required pre-fabricated standardized components in some situations. For these firms, transporting large ducts to the construction site and cutting them off saves the costs of materials and transfers the installation's responsibility to the installer.

- M&S in this case study prompted additional material costs due to splitting the structure of the systems, as well as subsequent need to additionally fix points and components. However, due to the fast construction processes (through the pre-fabrication of standardized MEP components, and the simple management system on the construction site) and the fast adoption process of the building during the operation phase, the additional material costs were not important. This means that the evaluation of the costs should not be based upon the benchmarking or mainly on average values, but should also consider other factors such as adoption possibilities and the fast construction process. Unfortunately, precise data about the costs and construction time could not be obtained.

It can be concluded that:

- Managing interfaces between project participants is necessary. The availability of an atmosphere promoting coordination and cooperation can be achieved through the use of LPS and the availability of incentives. LPS will help to manage the strong need for coordinating the structure of the building systems and will allow the participants to be more active by applying the rules of LPS (e.g. pull system to manage interfaces, integrating of construction team). On this project, work was achieved with "force", which may restrict the innovations and honesty and increase iterations during design.

- Developing tools to support the installation process of the modularized and standardized structure of MEP systems should be undertaken at the right point in time to assure their availability on the construction site at the right time.

- Analyzing the point in time to integrate the installer to develop these installation tools or introduce installation knowledge is essential to manage knowledge flow during design. This integration of the knowledge of construction and installation was a special case on this project, although it should be available in future projects with modularized and standardized structures.

Figure 6.8 shows the importance of integration of construction and installation knowledge during the implementation of M&S.

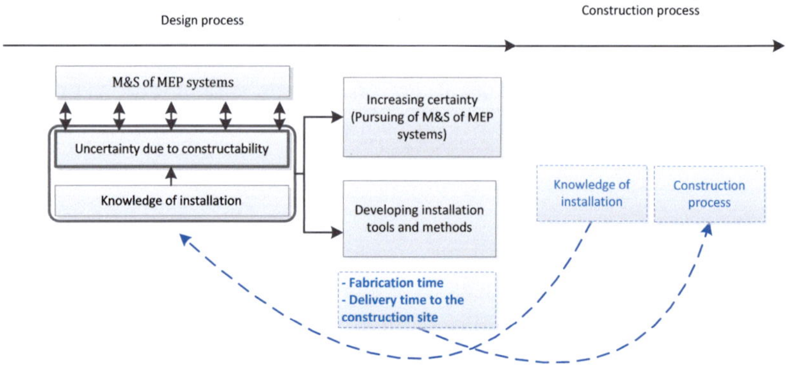

Figure 6.8: Importance of integrating installation knowledge during M&S processes

Table 6.6 shows aspects of implementation and potential of improvement of this case study depending on lean perspectives.

Table 6.6: Aspects of implementation and potential of improvement (case study 4)

Aspects of Implementation	Aspects of Improvement
What: Developing installation supports during implementation	
Why: Installer (he was also the sub-contractor) of heating and plumbing was integrated in design and he developed installation supports to realize the developed modularized and standardized architecture of MEP systems.	
Findings: Integration of the knowledge of installation was important for: (1) convincing the designer of the constructability of the system; and (2) developing installation supports to be delivered on time to the construction site and facilitating the construction process.	Achieving such integration should be assured in future projects with modularized and standardized structures.
The owner is worried about warranty claims during the construction and operation of the modularized and standardized structures.	Developing installation supports and analyzing installation conditions can be achieved by using trade-off curves during the implementation of M&S to prove the performance of the developed system.
Dissatisfaction with the modularized and standardized structure due to changing the way in which the designers should work.	Incentive system is required to improve cooperation and create satisfaction among the participants to use their innovative capacities.
M&S caused additional installation costs but there was no clear analysis about these costs and their impacts on implementation during the design.	Cost should drive the implementation in the design to achieve end-customer value. Using TVD will allow identifying opportunities to reduce costs.

6.5 Results of Application of the modeled methodology on the construction site: Case study 5

6.5.1 Heating system and pre-fabrication: case study 5-1

In this project, the application of the modular design strongly reduced the construction complexity. Furthermore, M&S in design enabled the use of a different installation process for the heating system. Originally, it was planned to weld hot-water pipes in place, although standardization of structures enabled the off-site pre-fabrication of pipe systems. This new production process was expected to reduce construction costs and time, as well as improving the quality of the construction and checking process at the construction site. Unfortunately, the savings in costs and time could not be realized. The now faster installation process could not be executed in a continuous flow, but rather in a stop-and-go manner, because its speed was not aligned with the speed of the other installation processes. Figure 6.9 explains the planned and executed process of installation of the heating system.

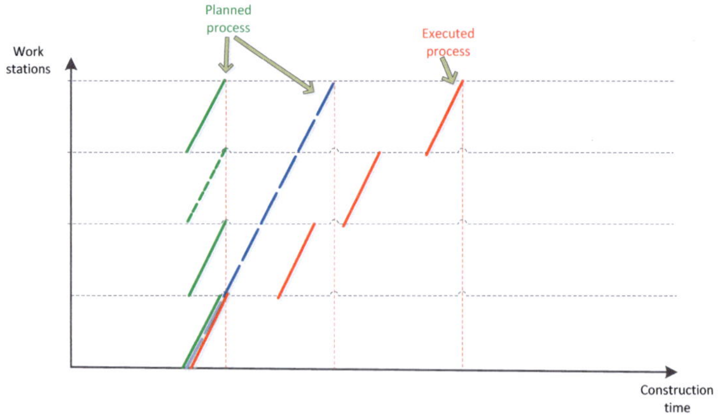

Figure 6.9: Planned and executed heating system process

Figure 6.10 shows above the pre-fabricated materials and pipes of heating (above) and the final product in place (below).

Figure 6.10: Pre-packaged materials for MEP-system installation (above); installed MEP systems (below) (source: Digitales Bauen, Karlsruhe)

Other MEP systems were not pre-fabricated off-site, but rather on the construction site, because the sub-contractors did not have the possibility/willingness for pre-fabrication, arguing that the costs would be higher if they produced and delivered small standardized parts.

6.5.1.1 Discussion

The construction firms did not have sufficient understanding about the goals of pre-fabrication and standardization according to the developed design structure and they wanted to optimize their costs locally through producing other sizes of the components on the construction site.

M&S of one system (heating system) was not sufficient to harvest the expected improvement in construction process, because the improvement concept depends on defining modules of all MEP systems or for every system. The modules should be defined during the design process.

It can be concluded that the causes of non-realization of the developed structure in the design are the reluctance to change the construction system and construction management, as well as the absence of communication between designer and construction teams to understand the system and its goals. A further important cause of non-realization – according to interviews with the designer and construction firms – is the decision-making strategy, whereby the decision is mostly made by the "account department", which receives the bills from the sub-contractor. The account department does not know about the benefits of the system and is only interested in the costs introduced from the construction teams/firms. One way to solve this problem is to integrate sub-contractors and the accounts department in the design to facilitate understanding of the system and allowable costs and align their values with this system. However, this requires other types of contracts like an integrated form of agreement. Further research will be necessary to discover the required contractual aspects.

Table 6.7 shows aspects of the implementation and potential of improvement of this case study depending on lean perspectives.

Table 6.7: Aspects of implementation and potential of improvement (case study 5-1)

Aspects of Implementation	Aspects of Improvement
What: M&S were implemented in design with conditions of subsequent bidding and competition system for construction.	
Why: M&S are values for the owner. According to the bidding process used, the construction firm should realize the developed structure in the design.	
Findings: M&S enable "tact" during construction. However, there was a reluctance to change the construction system and construction management.	Integration of construction firms (construction partners) early (during implementation of M&S in design) to understand and agree on the system and its benefits.
M&S of one system (heating system) was not sufficient to harvest the expected improvement in the construction process.	Global optimization is required to achieve the improvement in construction process.
Decision-making during construction is mostly made by the "account department", which receives bills from the sub-contractor. The account department accepted the offers of the sub-contractor to reduce the costs.	Introducing transparent knowledge about construction costs during implementation of M&S in design through integration of the required stakeholder to avoid later changes.

6.5.2 Sprinkler system: case study 5-2

In this case study, the type of the building is a healthcare company that operates under two divisions: pharmaceuticals and diagnostics. The previous design methodology of M&S was used during the design. The construction manager was interviewed to inspect the process on the construction site.

In this project, some requirements are defined: pre-fabrication for MEP elements is an obligation and most of production activities have to be conducted off-site.

According to the interview with the construction manager, the construction firms did not always hold the suggested sizes of the MEP elements that resulted from the modular design. Because many subcontractors existed, these changes in sizes could be made differently.

According to the perspective of the construction manager, there is a difference between modularization and standardization in design and what can be made on the construction site due to different perspectives of the designer and construction firms regarding costs. This occurs because the standard elements defined by the designer could cause more costs in comparison to the standard elements that can be made and transported by the construction firms, because the construction firms and their supplier (or fabricator) have pre-defined standard elements, mostly in larger sizes than they have to transport and install according to M&S developed through design. Making new standardized sizes for the MEP elements will cost more and cause a loss of material during pre-fabrication (on site or off-site), according to the construction manager. Furthermore, the pre-fabrication is sometimes not possible for the sub-contractor, because they can save more money by working directly on the construction site. According to the perspective of the construction manager, the difference between modularized and standardized systems developed in the design and a workshop plan is normal and must be accepted.

The realization of M&S could cause additional costs, resulting from: (1) more work in the factory to produce small standard elements; (2) more material due to producing small standard materials from large standard material; and (3) greater efforts due to more joining points.

The construction manager claimed that the MEP elements' sizes developed during the design only considered the optimization of utilization aspects but not the production process on the construction site sufficiently.

The construction manager explained that several problems emerged during the bidding phase:

The structure of MEP systems was modularized and standardized and accompanied with lists of components that should be produced and installed. In the bidding system, there were constraints for the construction firm/sub-contractor to accept the required supposed lists of components and a lower price was introduced for alternative prefabricated standardized lengths of pips. However, if the owner wishes to realize the developed standardized and modularized MEP components, he will demand an analysis about what he could lose if he accepts the offer of the sub-contractor. According to the interview, the savings of the designed modularized and standardized systems are theoretical and unrealistic.

One example of changes in the sizes of MEP components in this project is that the sub-contractor offered high cost savings if he could transport and install sprinkler pipes of six meters in length, which was much longer than defined according to the modularized and standardized structure. The decision to accept this change only required inspection of the related logistics of the pipes from the workshop to their installation places. The length of the pipes described in the bidding is mostly a global length for the piping or a maximum length for the units of piping. Typically, there is no obligatory precise description of the components. If there is a precise length for the units of piping, changing the length subsequently needs approval and analysis is required according to cost and logistic aspects. Although, the suggestion to use long pipes was refused and the decision was made only due to logistic aspects, without considering the compatibility with the construction process of other MEP systems. The potential of global optimization of the construction process of MEP systems and factors of flexibility against possible changes in the design (made by simply de-installing parts of pipes) were not considered as criteria in making this decision.

According to the perspective of the construction manager, the designer must know more about the standard available sizes of elements to reduce the conflicts between the designer and sub-contractor.

6.5.2.1 Discussion

The conflicts - which are not considered at the right time - cause iterations and re-work, because during the construction process the designer must inspect the new supposed lists of components and the resulting required changes in the other dependent MEP systems. Waiting on the construction site for the new inspections and decisions caused waste and interruptions in the workflow. In this situation (the sprinkler system) and according to the designer, although there were no required changes to other dependent MEP systems, such changes could cause a re-design or re-structuring of some systems or components. The construction firm will mostly quote a high price to produce and install the modularized and standardized structures, where factors of investments and methods of cost calculation and distribution are essential to cope with additional costs. The separation of the departments and the hand-offs cause a significant loss of information, which strongly affects the decision-making process.

The modularized and standardized structure aimed to define standardized modules comprising pre-fabricated elements that can be prepackaged and transported to their installation place on the construction site, whereby they can be installed continuously or in parallel. The construction teams consider cost factors that do not consider improving flexibility against design changes and productivity on construction site, which restricted the realization of the modularized and standardized structure. Figure 6.11 explains how different perspectives of the design teams and construction teams led to reducing M&S of MEP systems.

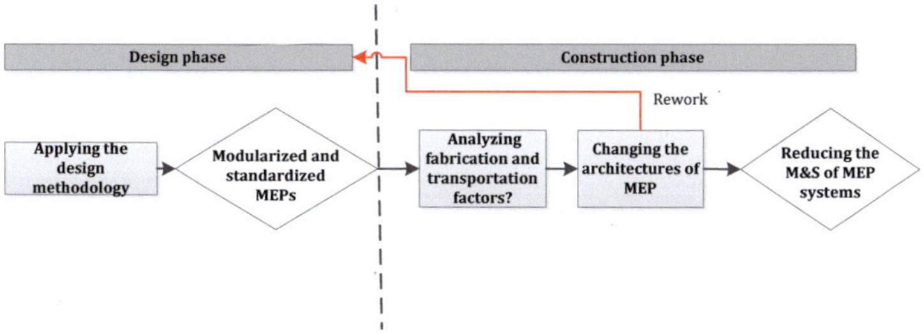

Figure 6.11: Different perspectives of design and construction teams lead to reduce M&S of MEP systems and re-work

Making other sizes of the components (in this situation, pipes) allowed pre-fabrication, albeit not in compatibility with the installation process of the other systems. Moreover, the benefits of flexibility were reduced.

A module of MEP systems mostly contains all required MEP elements, which can be used in other spaces of the building and can be easily replaced. This shows the importance of M&S against possible changes in design, whereby making standardization in another way (where the interfaces or the boundaries of the modules are made differently) may reduce the potential of flexibility against design change. Precise analysis about the extent to which changing certain elements can affect the potential of flexibility against design change and the improvement of construction process will help the decision-maker in design to decide about conflicts from the perspectives of the design and construction teams.

The decision-maker during construction has little understanding about the importance of maintaining the boundary of the modules as they were defined during design as an effective structure if changes in design occur, as well as to improve the construction process (e.g. through pre-fabrication and takt time planning). Moreover, trade-off analysis conducted to make the decision of changing the lengths of the pipes did not consider all factors; for example, the stability of the workflow of the MEP installation process and savings in construction time. On the other hand, the design team did not consider the perspective of the construction team and their capacity, because they were not known in the design phase and they were not allowed to provide feedback about their values and perspectives.

This case study shows how late changes could affect design factors that should be re-checked to assure the quality of design. In turn, this not only causes re-work, but also a reduction of M&S of MEPs. The case study shows how project participants tend to engage in local optimization. For the construction process, not aligning with the other MEP systems hinders the stability of workflow of MEP systems on the construction site (and subsequently the savings in construction time), which can be achieved by following up the structured systems. By making such changes, logistic and installation processes can no longer be made according to the defined modules in design and it will not be possible to realize the parallel installation.

Typically, the designer has restricted knowledge about the perspective of the sub-contractor. The decision made by the owner and his representatives is to follow savings in costs suggested by the sub-contractor. However, these costs should be compared with the other components of the costs, including factors of flexibility, the pre-fabrication of standardized components and the stability of the construction process.

It can be concluded that through the design process, two elements have not been sufficiently inspected – namely the constructability and cost evaluation of the modularized and standardized systems – and that conflicts are caused due to a separation between design and construction and insufficient mutual understanding of the perspectives of different stakeholders. The construction teams thought about the installation of the entire sprinkler system to achieve the required functions with reduced costs, although they did not consider the transportation and installation of modules of sprinkler systems with alignment with other MEP systems as they were defined by the design team. Therefore, throwing over the wall caused re-work and other types of wastes.

Table 6.8: Aspects of implementation and potential of improvement (case study 5-2)

Aspects of Implementation	Aspects of Improvement
What: M&S were implemented in design under conditions of subsequent bidding and competition system for construction.	
Why: M&S are values for the owner. According to the bidding process used, the construction firm should realize the developed structure in the design.	
Findings: The sub-contractor introduced a lower price for other pre-fabricated standardized lengths of pipes for the sprinkler system during the construction phase.	Waiting on the construction site for the new inspections and decisions caused waste, unnecessary iterations and interruptions in the workflow. Making such a decision at the right time (during implementation in design) should reduce iterations and waste.
The construction teams considered cost factors that disregarded total productivity improvement on the construction site. This restricted the realization of the modularized and standardized structure.	- Different perspectives of design teams and construction teams lead to reduce M&S of MEP systems (construction team was not known during design). - Tend to make local optimizations because the decision to deliver and install the sprinkler pipes did not consider other aspects. - The integration of a special supplier/sub-contractor who commits to realize the defined modularized and standardized systems early in the design process will reduce possible changes during construction.

It can be suggested that that integration of a special supplier/sub-contractor – who will commit to realize the defined modularized and standardized systems and benefit from them – early in the design process can reduce possible changes caused by integrating the supplier and construction firms (and their knowledge) later. Table 6.8 shows aspects of the implementation and potential of improvement of this case study depending on lean perspectives.

6.6 Case study 6: Design process: Challenges

In this case study, the type of the building is a "test bench" building. The goal of this case study is to observe elements of M&S during the design process, as well as elements of the thinking strategy that could hinder M&S of MEP systems: in other words, the goal is to identify the conflicts and challenges of implementation. Analysis was conducted through interviews and hearing to discussions during the design phase. Observations in this case study provided information related to the analyzed design methodology (M&S of MEP).

M&S of MEP systems are undertaken through developing of standard units of utilization and standardizing of the interfaces between MEP systems and the utilization of spaces to achieve flexibility in the utilization of spaces.

Thinking about standardizing MEPs' structures could take place first in the construction planning phase, which has many causes. According to interviews with the project participants - especially the MEP designer - standardization is the task of construction firms rather than themselves because it deals with the installation process of MEP systems' components. According to previous case studies, M&S of MEP components is a design and construction aspect and not only a construction aspect. Therefore, achieving this goal requires cooperation, not only between the designers, but also among all stakeholders that could be affected.

Figure 6.12 represents some elements of the design phase in this case study. The represented elements were chosen because they relate to M&S of MEP systems in design and construction planning.

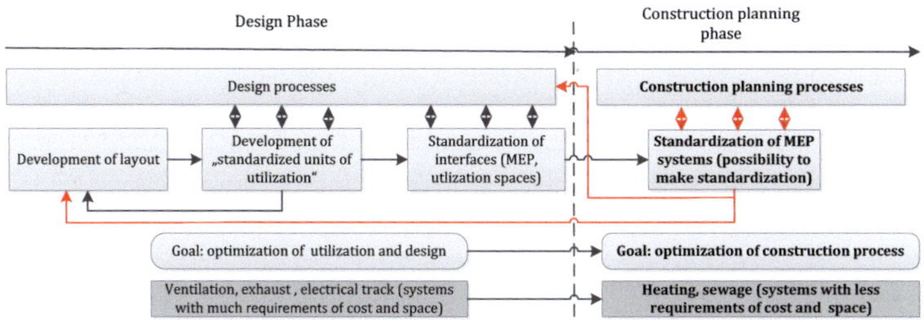

Figure 6.12: Design and construction planning processes

The represented elements of M&S in Figure 6.12 can be explained as follows:

1. Development of modules of utilization. These modules are developed by the user and they can be installed in standardized spaces. Another aspect of the utilization is standardizing the interfaces between MEP systems and the spaces of utilization to enable flexibility in utilization, i.e. different types of utilizations can be made in the same spaces. Flexibility of utilization is the core aim of M&S during the design process. The users were intensively integrated during the design process to define the interfaces with the MEP systems.

2. Modularization concept (MEP systems): the concept of modularization is made by oversizing interfaces between MEP and spaces of utilization (standardizing of these interfaces) to obtain the same interfaces in different spaces for different types of utilization. Therefore, the integration of the user in design is essential to define suitable oversizing of the interfaces. However, the communication pattern between the MEP designer, the user and other participants needs to be

observed from the perspective of information flow between them to define types of waste and reworks. However, this was not possible in this study. The standardization of interfaces can also enable reduction in costs of operation by allowing only the initial partial operating units within the spaces and joining more operating units in the future, which will not require changing the basic piping system. The cause of joining new operating elements in the future are considered in the current design through oversizing interfaces and planning free spaces for the new operating systems in the future.

3. Not all MEP systems are considered during the design phase: the exhaust and ventilation are normally needed by a test bench with large diameters of piping and subsequently for more places in the technique floor. The heating system and waste water system are designed later in the construction planning phase. The reason for this thinking is that the large pipes need more spaces and greater costs, which need to be estimated to optimize the costs (from the perspective of the designer), as well as the low cost of heating and waste water systems - in comparison to ventilation and exhaust - could be added later without problems.

There are priorities in the design process, which the MEP designer defined according to the cost factors (such as the electric traction) and design requirement factors (like the waste water system). This means that the systems or parts of the MEP systems that have more costs are designed first, followed later by the systems or parts that cost less. This thinking of priorities is a challenge to implement M&S of MEP systems, which depends on defining modules - which are volumetric parts - and designing all the required MEP systems to obtain the required utilization (function) in this volumetric part.

In an interview with the users, the importance of integrating the user through design could be recognized. The users have developed standardized units of utilization to be installed in standardized spaces, and they defined all their requirements in a book before starting with the design process; however, it will be difficult for the designer to ascertain the required information of utilization at the right time during the design. Therefore, integrating the users is important to ensure that the required information about interfaces between spaces of utilization and MEPs are available at the right time. If the structure of the MEP systems affects the utilization's aspects, integrating users during the design is also important for the negotiation process in terms of knowing how the users could be affected by M&S of MEP systems, as well as what changes they could accept to improve M&S of MEPs, especially given that users provide some information like intervals and not determined values. Regarding the standardization of MEP structure, one user in this case study mentioned that the standardization of MEP systems is not important compared with the optimization of the utilization.

Figure 6.13 represents a section of the detailed design of the ventilation system. The distance between the outlets on the main route is not standardized, and the components of the sub-routes are different (different lengths and different bend angles).

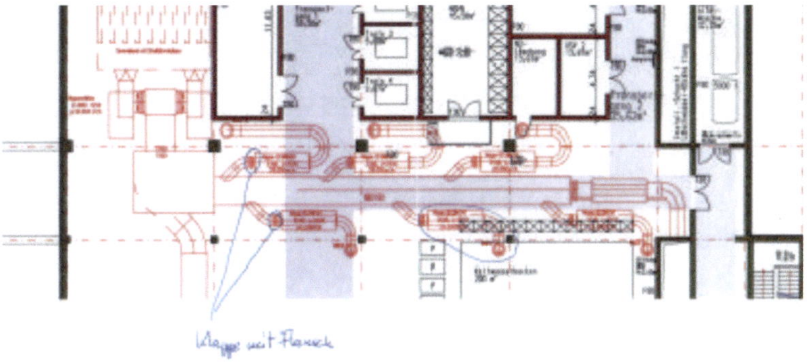

Figure 6.13: Variety of components for the ventilation system

The causes of the variety of the components are geometrical and structural constraints, which can be reduced or eliminated through communication with the specialists (designer) to find suitable solutions. However, defining standardization as a value is a prerequisite for successful communication between the customers. According to the interviews, the designers are not obligated to undertake this standardization, which requires greater effort and coordination.

6.6.1 Discussion

According to the studied methodology (case study 1), flexibility can be achieved through M&S of spaces, utilization and MEP structures, where M&S of MEPs lead to significant improvement in the construction process through mass production and reducing the variety of components. In this case study, there is no focus on the construction process of MEPs, and flexibility is achieved through developing standardized spaces for modules of utilization and the over-design of interfaces between spaces of utilization and MEPs. Improving the construction process through M&S of MEPs structure can be aimed to cope with additional planning and material costs (due to flexibility). However, the optimization of MEP structure is achieved later in the construction planning phase because it needs alignments with construction firms, which constrains possibilities of standardization or causes additional iterations in design (if standardization is desired), because it may require adjustments in geometrical, structural and utilization aspects. From the perspective of the designer, if other participants like construction firms participate, they will not offer their evaluation faithfully.

Applying M&S concepts requires considering all MEP systems. This will help improve the construction process through reducing the variety of components and enabling takt time planning for the construction process of MEP systems. The challenges for M&S are as follows: (1) the standardization of heating and waste water will be very difficult because they must be placed in the free available space, given that the ventilation and exhaust have already been placed and changing them will cause numerous iterations (more cost and time); and (2) this separation leads to less conflict in the design process than in the situation whereby all systems are integrated (interdependencies within MEP systems), although it causes less alignment of customer values, represented by the goal of all project participants modularizing and standardizing the MEP systems. For example, the increased conflicts make the designer think more about all possible situations of positioning the piping, resulting in more conflicts in design, but greater alignment of values. This shows the importance of applying LPS to integrate all project participants and considering interdependencies at the right time through pull planning.

The specialists were encouraged to make sub-optimization and set priorities based upon their perspective of what is important. This thinking is a challenge to M&S that requires a management system to counteract this way of thinking. LPS focus on defining the activities near in time to their performance considering design criteria to improve value generation and workflow between all specialists, as well as avoiding sub-optimization and priorities-based thinking (Ballard 2000c).

Table 6.9 shows aspects of the implementation and potential of improvement of this case study depending on lean perspectives.

Table 6.9: Aspects of implementation and potential of improvement (case study 6-1)

Aspects of Implementation	Aspects of Improvement
What: M&S of spaces and utilizations is achieved. LPS was used to manage implementation in design.	
Why: Flexibility of spaces and utilization is defined as value for the end-customer. M&S of MEP's structures were not thought about and considered as tasks for construction teams.	
Findings: Integration of the user through implementation improves performance during implementation of M&S.	The user thought that aspects of utilization are the most important thing, and that other aspects of the design should be adjusted to optimize the utilization.
M&S of the structures of MEP systems were not made due to the perspective that this is the task of the construction firm and this will cause more efforts in design, which is not considered by HOAI.	– Integrating construction firms to participate in modularizing and standardizing the structure of MEP systems can achieve improvements in the construction process and construction costs. This could be used by evaluating the costs. – Using other types of contracts and payments for designers who do not depend on HOAI to motivate them to cooperate transparently during the implementation.
Interests of design and construction are not aligned due to used types of contracts.	Using other types of contracts such as IPD to align the interests of the stakeholder, which is important for a transparent cooperation to achieve M&S of MEP.
Sequential implementation in the design can reduce the potentials of M&S of MEP systems and cause unnecessary iterations.	Integrated development during the design and using tools such BIM facilitate the integrated development and reduce iterations.

6.6.2 Standardization of structural system

The goal of introducing this example is to explain the thinking strategy of the designer and construction firm if standardization will be undertaken. This example shows the dependency between standardization as a method to improve the construction process and the design process. The existence of such a dependency causes conflicts in the design process, including design conflicts, cost conflicts and satisfaction conflicts. The analysis includes answering the question: why do such conflicts occur?

Standardization was thought for the structural system in this project as a method to improve its construction process regarding crane movement on the construction site and achieve faster construction process. For this reason, the owner invited a construction firm during the design to optimize the construction process. The construction firms standardized the structural system to improve the construction process, which – according to an interview with the MEP designer – restricts the MEP design and causes a re-work for the structural system and architecture in the under floor. The iteration in the MEP design will be small because they are not detailed until the point of time of standardizing the structural system (at the beginning of construction planning phase). This shows two aspects: the dependencies between structural planning and MEP planning, which cause iteration in design, as well as the possibility to undertake standardization by integrating the construction firms.

Due to standardizing the structural system, the construction firm will be three weeks faster. However, the standardization will increase other costs like materials, according to the MEP designer. Although the construction firm has not introduced any cost evaluation of its system, the system was accepted because the construction process will be faster, which is very important for the owner. The MEP designer was not satisfied about the standardization of the structural system – suggested by the construction firms – and he considered that the benefit of this system will only return to the construction firms to improve their construction process. MEP designers also have the perspective that integrating the construction firm will certainly lead to more costs because by breaking the system of competition, the construction firm will exploit the situation of a lack of competitors and subsequently significantly increase the construction prices. Moreover, he considers that this type of exploitation is a normal human behavior.

Through listening to discussions and asking about this subject, to standardize the structural system, the dependencies between the construction activities and MEP design have been identified. These activities include the possible fixation points and adherence for the crane. These points are important to enable certain construction activities that resulted due to the standardization of structure or transporting and installing pre-fabricated elements. The standardization of structure enables the pre-fabrication of structural elements, meaning that the transportation of the structural elements to their places in the building will be undertaken by a crane, which needs fixation points and movement areas, thus affecting the MEP design, e.g. in terms of where to place the pipes. This interdependency led to a communication process to define the possibilities and make a decision about this. However, tracking the communication patterns was not possible in this case study.

From the perspective of the MEP designer, the new plan is only beneficial for the construction firm and it will cost more for the owner, who did not conduct a sufficient analysis and trusted the construction firm directly.

Another challenge that hinders the benefits of standardization is that the evaluation of different alternatives (standardized or not standardized) is undertaken by the calculator, who have no experience in evaluating the alternatives according to criteria regarding to the production method. Therefore, for example, the alternatives that may have high benefits on the construction site will not be evaluated correctly, because the people with the required experience are not involved in the calculation of the costs.

6.6.2.1 Discussion

The new design of the structural system requires a new design for the MEP systems, which will be restricted due to the standardization of the structural system. There was a great conflict between the participants about the decision taken. According to the MEP designer, the benefits will only improve the construction process and increase the profit of the construction firm. The possible cause for this conflict is that there are many concerns among many participants, and the decision to standardize the structural system was made due to cooperation between the owner and the construction firm to achieve a faster construction process. In turn, this led to design solutions that are not understood by the designer. The construction firm may exploit the no-competition system to increase their own profit. Thus, there was no transparency among the different project teams, whereby every team thought that the others introduce solutions for their own benefits. Moreover, the owner - who wanted to achieve a faster construction process - forced the designer to accept the solution of construction firms, which was not understood by the designer.

The improvement in the construction process would not be evaluated according to the calculation system, which mostly depends on middle values of different types of costs. Thus, the real improvement in the construction process was not calculated.

In this case study, the challenges to implement M&S of MEP systems in the design can be explained as follows, benefiting from the situation of standardizing the structural system: (1) challenges due to increased design interdependencies; (2) challenges due to separation between the design and construction planning and the construction phase (bid and competition system); (3) the calculation system; (4) different concerns of the project partners; and (5) the need for transparency among project teams by evaluating the costs and decision-making by the owner. Table 6.10 shows aspects of the implementation and potential of improvement of this case study depending on lean perspectives.

Table 6.10: Aspects of implementation and potential of improvement (case study 6-2)

Aspects of Implementation	Aspects of Improvement
What: Standardization of structural system in a late phase during the design to improve the construction process and reduce the construction time.	
Why: To inspect the late implementation of standardization.	
Findings: MEP designers were not satisfied with the system because they must adjust their systems to achieve the standardized structural system.	Hand-offs of the developed standardized structural system led to dissatisfaction, which will affect the cooperation and transparency of MEP designers.
Possibilities of rising construction prices because the construction firm will not have competitors.	Transparency is required to avoid increased prices using other types of contracts such as IPD. Calculation of costs should be based upon precise knowledge.
Late implementation of standardization caused iterations.	Early implementation of standardization should be assured to reduce iterations.

6.7 Case study 7: Types of wastes during design

This case study was based upon a current project. The type of building is a healthcare company that operates under two divisions: pharmaceuticals and diagnostics. The previous design methodology of M&S (case study 1) was used during the design. We interviewed the design teams through multiple telephone calls.

According to the interviews, the goal of the implementation of M&S was to achieve flexibility in the utilization of the building, e.g. generating large spaces of utilizations from small spaces, changing the type of utilization in some spaces (e.g. workshops can be later used as offices).

The following challenges and conflicts were referred to by the design teams:

1. Most conflicts take place with the architects, who like to create variety. However, this variety does not necessarily increase the end-customer value.
2. One important challenge is the need to increase detailing in the early design to analyze the possibility of the modularization and standardization of MEPs. This challenge occurs because not all participants are known in the early design.
3. Another challenge appears in spaces where there are different types of utilization, and breaking up the spaces into small modules to increase M&S will be uneconomical, according to the perspective of the MEP project leader. However, the additional costs emerging due to M&S could not be identified in terms of numbers or percentages by the team of the design phase.
4. There are different priorities for different participants, and the information for decision-making available to each participant is not the same, which causes iterations and multiple adjustments.
5. Some conflicts appeared because the modules reduce end-customer value regarding utilization issues. This means that M&S are not aligned to some requirements of utilization.
6. Design teams confirmed that integrating users in the implementation of M&S will not increase the potential of M&S

because the user defined their requirements, whereby the structure of the MEPs is not important for them (this is the same as in case study 6).

By discussing the unnecessary iterations with the project teams, they explain that the core cause of these iterations was the unavailability of required information at the right time; for example, defining the positions of the outlets depends on the defined boundaries of the modules. However, because the specialists of the ventilation system were assigned later, there were two possibilities: waiting for the specialist (this is a type of waste: waiting and waste of resources) or defining the boundary of the modules without alignment with the specialists of the ventilation system, which cause difficulties in designing ventilation system and iterations (between this process and the process of defining the boundary of the modules). Thus, defining the boundary of the modules not only depends on the requirements of the user (utilization), but also on aligning with the design of other systems, which could require redefining the requirements of utilizations if more costs or efforts could occur and the owner does not want to pay for them.

One participant in design - who was included in the first part of the project (he was not included in the engineering phase, where new teams was included) - referred to the main three factors that negatively influenced the implementation of M&S:

1. The different levels of knowledge among the participants led to multiple adjustments and iterations.

2. Not all participants were available to provide required information at the right time, which resulted in waiting for this information or making decisions with incomplete information. The unavailability of the required information was either because the participants existed in different places or because the stakeholders were not known given that they were not yet assigned to the project.

3. The decisions were made very slowly by the owner, and he was unable to exactly define the flexibility concept early (i.e. sectors and extent of flexibility in the building).

The next interview was about identifying of types of waste during the implementation of M&S that are caused by the previous challenges. The following types of wastes were defined in relationship to the implementation of M&S, as seen in Table 6.11:

- The complex generation of required information: for example, information about utilizations are necessary in the early phases. However, the users are not known in this phase, and communication with them is not allowed or restricted.

- Multiple adjustments to the definition of the modules: this is caused due to the sequential implementation and unavailability of detailed information in the early phases. The unavailability of detailed information about the modules (Information about technical systems (MEP systems)) was caused because the designer of MEP systems were not integrated in the process of defining the modules. This led to restrictions in MEP design and redefining modules later in the engineering phase, where the MEP designer was available.

- Too much detail, which is not needed.

Table 6.11: Types of wastes and causes: Case study 7

Type of waste during implementation	Cause
1. Complex generation of required information.	• Different levels of knowledge by the project teams cause unshared information. • User are not known. • Communication with user is not allowed directly. • Project participants work in different places. • The decisions were made very slowly by the owner.
2. Multiple adjustments for the module size (module boundaries) and their interfaces.	• Sources of information are not consistent. • Decision-maker are changed. • Priorities changed . • Different levels of knowledge by the project teams cause unshared information . • User are not known. • Direct communication with user is not allowed.
3. Too much detail.	Incorrect focus on not required information.

6.7.1 Discussion

It is clear that M&S need greater efforts during development due to more interfaces within every level (geometry, utilization, component (engineering)) and between the levels. Therefore, more information should be available in the early phases and integrated design should be implemented to improve learning. Integrating active players who can deliver required information in the process of defining modules is necessary. These players are the users and the MEP designers. The late integration of these players leads to long learning cycles, which cause long iterations, as explained in Figure 6.14.

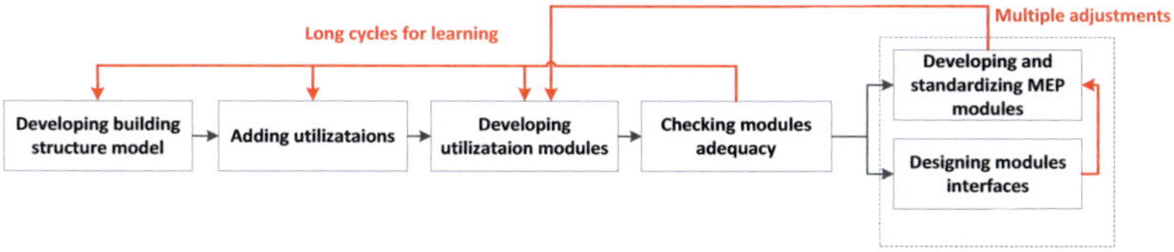

Figure 6.14: Long cycles of learning and multiple adjustments during the implementation of M&S

The long iterations are caused due to a sequential strategy in managing design process, whereby not all stakeholders participate in the process of defining the modules. Thus, defining modules from one perspective and then handing them off to other participants to check the alignment can be described as a "hard system", which is simply not lean and causes iterations during the implementation.

The multiple adjustments to the modules and interfaces of the modules are also necessary due to sequential implementation, late assigning of the active player and not detailing the design at the right time, which leads to neglecting important information (for example important interfaces). Pull planning and LPS can significantly improve this process.

More information about the utilization should be available in the early phases and they require fast decisions from the owner, which can be improved by integrating the designer and those responsible for M&S in the early phases of design, which means engaging greater efforts in the early phases to reduce late iterations. Defining a flexibility concept (sectors and extent) as a goal was achieved by the owner and his consulting and this concept were handed off to the other project participants, who make their tasks in creating the goal of the owner. This hand-off caused certain iterations and a lack of complete understanding of the concept.

The implementation process of M&S in this project can be described as follows:

1. The development of modules was undertaken by a responsible team, who had the information from teams that developed geometry and defined utilizations (sequential development, hand-offs).

2. If adjustments were necessary, the responsible team for conducting M&S were unable to do this directly due to the long waiting time for decisions that should be made by the owner or analyses conducted by the other teams, who could have had another level of knowledge given that they were not included before (for example, new teams are assigned). In this situation, the responsible team for undertaking M&S can:

   - Make adjustments without other teams, which could lead to their later rejection by other teams that were not included in this process, as they have other possibilities or perspectives to make these adjustments.

   - Alternatively not make adjustments, which leads to reduced potentials of M&S with no impact on the owner or end-customer value.

3. Developing standardized modules of MEP occurs based upon developed geometrical and utilization modules, whereby having been handed off to the MEP designer, he could refuse them for many causes, such as the following:

   - The developed modules do not allow achieving the required performance.

   - They have other alternatives for modules of MEP systems, although they require many adjustments of geometry and utilizations.

- They do not like developing the MEP systems in another way; therefore, they can argue that this is not possible or it costs more money.

The user (and other design participants) should participate actively to achieve M&S through assuring that all participants have the same goal (achieving of M&S with alignment to quality).

The economic aspects - which could not be precisely defined by the project participants - should be analyzed and based upon knowledge of different affecting aspects and from different perspectives (e.g. using trade-off curves). One of the economic aspects that emerged in this case study is the increase in costs when breaking down the modules into smaller modules. Here, trade-off curves can be used to analyze the increased costs and the trade-offs with other factors (like flexibility, increase of standardization). Developing the trade-offs should be undertaken by the design teams to improve negotiations and the decision-making process.

Different types of waste are caused due to engaging the participants in different places or not assigning the required specialist at the proper time. Integrated types of contracts and co-location (large room) will reduce the waste and improve learning. This case study shows that the traditional management system leads to more iterations and a slow learning process during the implementation of M&S. It can be concluded that despite early thinking, there were conflicts and dissatisfaction during the implementation. Thus, early thinking is not sufficient because sequential implementation and hand-offs lead to long cycles for the learning process. Improvement in the implementation process of M&S based upon a lean perspective depends on improving the learning process through short learning cycles, which can be achieved through pull planning and LPS. Other lean tools like set-based design should also be inspected in detailed case studies in future research. Table 6.12 shows aspects of the implementation and potential of improvement of this case study depending on lean perspectives.

Table 6.12: Aspects of implementation and potential of improvement (case study 7)

Aspects of Implementation	Aspects of Improvement
What: Analysis of types of waste during the implementation of M&S of MEP systems.	
Why: Types and causes of waste can provide information about the reliability of the implementation.	
Findings: The implementation process includes many types of wastes due to the management system used. Types of waste can be classified under two main types: hand-offs and scatter.	Integrated project development will reduce the defined types of waste and assure transparency and thus improve the reliability of the implementation.
Increased efforts during the implementation increase the costs due to the need to align the customer values. The increased costs are the cause behind not aligning all customer values, which requires more efforts. More costs for the alignment are caused because payments for the designer were based upon time they spend on work.	• The increased costs can be better analyzed and balanced by integrating lifecycle cost and benefits (using TVD) • Not aligning all customer values will cause iterations and possibly missed opportunities to maximize customer value.
Defining the right detailing of design at the right time during the implementation is a challenge, which caused iterations.	Measuring the reliability of planning in relation to the modularized and standardized structure can help participants to define required design factors through a continuous improvement process. This measurement can be made during LPS' meetings to ensure the participation of all stakeholder.
Commitment to implementation is required for a successful implementation.	LPS can be used to manage the production system of implementation where commitment is an essential part of LPS.

6.8 Challenges in implementation based upon the previous case studies

The implementation of M&S of MEP systems is a long process that begins in the early stages of the project, before proceeding to the engineering process and installations on the construction site. According to the previous case studies, the challenges could emerge in the design or/and construction phases. According to an implementer of M&S, in the current situation, if the possibility to modularize and standardize the MEP is 70-80%, the realized percentage is 10-20%. Figure 6.15 shows two scenarios of current implementation:

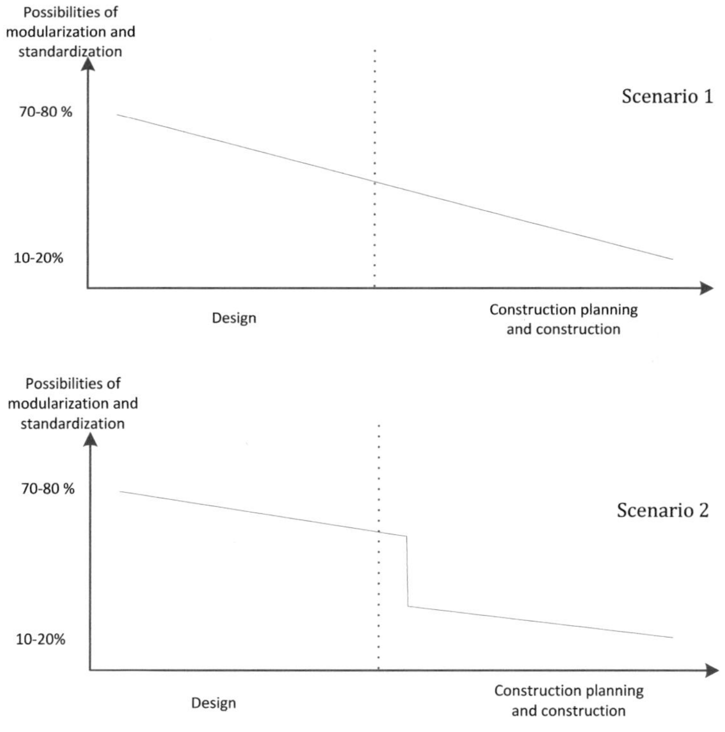

Figure 6.15: Reduced percentage of M&S of MEP systems in the design and construction phases - designer experience (scenario 1, scenario 2)

The first scenario shows that the challenges of implementation occur in the design and construction, while the second scenario shows that most challenges occur during the construction. The case studies and interviews assured this perspective, although the numbers depend on experience (or on his perspective) rather than statistical analysis.

In the next sections, the challenges will be summarized by classifying them into two groups: challenges during design and challenges during construction.

6.8.1 Challenges during design (design and prodution system design)

According to the interviews and case studies, there are many challenges during the design that could hinder or slow down the implementation of M&S of MEP systems, as outlined below:

1. Conflicts with other building systems: when performing M&S of MEP systems, the spatial and utilization requirements may need to be adapted. The conflicts are design conflicts and cost requirements. Design conflicts relate to dependencies between building design systems (architecture and structural systems), while cost conflicts arise when design conflicts must be solved through over-design; for example, where more material is necessary. As an example, the process of defining the modules is a challenge because it is an iterative process and it needs the alignment of the utilization, MEP systems and the layout. This means that it requires information from many project participants at the right time. Defining the dimension of the standard module (2,9 * 2,9 m) took a long time (one and a half years) in the project of case study 7.

2. The most important challenge in the design is designing on the level of defined modules rather than larger levels. This

way of design requires changing the way in which the design concepts are developed. Many more interfaces and constraints should be considered and coordinated. In the traditional design, the designer receives a layout from the architect to develop the MEP systems for the different spaces, although in the modular design all MEP systems (or part of them) should be developed for the modules. Moreover, the participants should also find design alternatives to standardize the MEP modules.

3. Coordination between MEP systems during the design: the conflicts between the systems could be increased due to restrictions caused by the modularized and standardized structure.

4. High detailing for MEP systems during design: this is a challenge because in the traditional way, this detailing is undertaken by construction firms who will install the systems and who come later after the design is completed.

5. Clear definitions of utilization requirements that need intensive integration of the user through the design to check the effects of M&S on utilization.

6. Conviction from the implementation: the designer should change the way in which they work, which is a major challenge. Furthermore, they must consider the potential benefits by changing their methods.

7. Same understanding and willingness to cooperate: because M&S of MEP systems make the structure of the MEP systems different, the designer should not only have an understanding to change their ways in designing and structuring, but also they must be able to cooperate. Having an incentive system and new types of contracts will help to achieve this.

8. Calculation system: the current calculation system is a simple system, which uses the average values of material, labor and transporting costs and disregards other types of value such as facilitating the installation process, the management system and flexibility according to design changes. Value-oriented evaluations must be undertaken by a professional calculator.

9. The HOAI payment system hinders the designer from engaging more efforts and flexibility of cooperation in the design, which is necessary during the implementation of M&S. According to HOAI, the project is divided to work phases, whereby payment for the designer's work is defined according to the construction sum (for example, installation of the ventilation system). Modularization and standardization are not included in the work phases of HOAI and they need greater efforts from the designer; for example, the designer must not develop a heating system for a room or space as in the traditional design, although he should also think about how the spaces of the building can be flexible by modularizing the systems, as well as how to standardize the developed modules. This require much greater effort in comparison to the traditional design, which should be recognized and considered by calculating payments of the designer.

10. Production control is necessary due to many existing interfaces between the project participants. For example, committing the MEP designer to maintain the boundary of the modules during design is essential and needs controlling.

11. Different priorities for different participants affect the decision-making process during implementation, making it centralized (participants should always wait for decisions made centralized by the owner)

12. Proving the more value achieved by the modularization and standardization of MEP systems for the owner. Introducing such a proof will motivate the owner to accept applying M&S because he decides what he wants to pay. One way to achieve this is to conduct a trade-off analysis during the implementation. However, this not only requires knowledge about additional material costs but also design planning costs, flexibility costs, operation costs and design change costs. Decision-making should be made according to allowable costs.

13. Choosing the suitable detailing of the building components in every design phase: this challenge was defined by the project teams of the case study 7, whereby the detailing affects the required information that should be available when needed. This is linked to organization structures and integrated development (for example, co-location).

6.8.2 Challenges during construction

1. During construction processes, there are typically new partners who can achieve reduced construction costs according to a system of cost calculation used. The integration of special suppliers early in the design making a commitment to realize M&S will help to realize the developed modularized and standardized structure of MEP systems, whereas integrating them late may lead to changes depending on the cost calculation system and not sufficiently understanding the system.

2. The risk resulting from new configurations and installation processes. Decomposition in the structure of MEP leads to new installation systems with many interfaces and connections, which could cause warranty claims.

The implementation of M&S is difficult because the challenges are not restricted to narrow levels such as design methodology, but rather they extend in connection with the broader organizational levels.

The challenges can be classified into two types: hard factors and soft factors. Soft factors do not relate directly to the building itself (as contract and organization issues), unlike hard factors (design factors). In this research, it is explained how these two types of challenges are interrelated. The implementation under another type of contract that profits design and construction teams and incentivizes designer and construction teams (contractors, supplier and fabricators) to cooperate early in design will significantly reduce the challenges.

The following section introduces an implementation model of M&S based upon the knowledge obtained from the previous case studies to define challenges, as well as lean concepts and tools as a way to reduce the challenges.

7 Developing a management system to implement M&S of MEP systems

The required knowledge to implement M&S of MEP systems concerns how to make and facilitate these processes. Performing M&S requires a methodology that defines the main processes of implementation, whereby facilitating relates to the management of knowledge and providing conditions of cooperation.

Three main challenges will be discussed in this section:

1. M&S processes are iterative evolving processes that require knowledge from many participants.
2. Uncertainty by the participants and owners about manufacturability and warranty claims.
3. Unreliability of planning of the implementation due to many types of waste.

7.1 M&S processes are iterative evolving processes

According to the literature review and case studies, the challenges of the implementation of M&S appear due to design interdependencies as well as management and organizational aspects. The previous types of challenges are interconnected and their nature leads to suggesting lean tools and methods as a strategy to achieve M&S of MEP systems, which will be explained.

The following diagram in Figure 7.1 shows general relationships between the implementation of M&S and other activities during the design process. The diagram is based upon the modeled methodology in case study 1. It explains the importance of considering the implementation of M&S as a part of the design process where dependencies exist between processes of the implementation and other design activities.

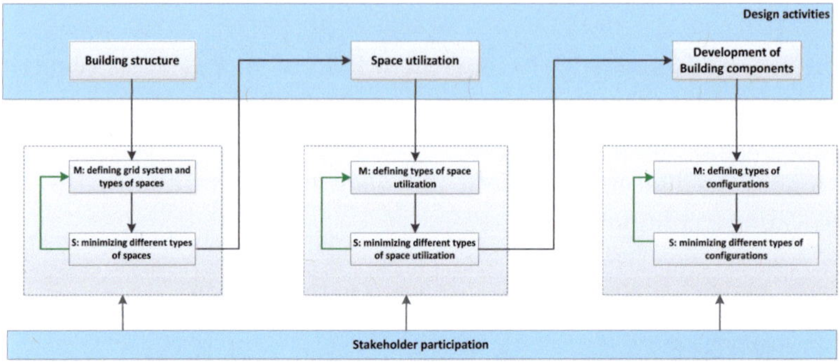

Figure 7.1: Implementation of M&S processes during design. M: Modularization; S: Standardization

Early thinking about M&S of MEP systems during design must be ensured. Pre-fabrication and installation should be considered in the engineering process to ensure constructability factors, as seen in Figure 7.2.

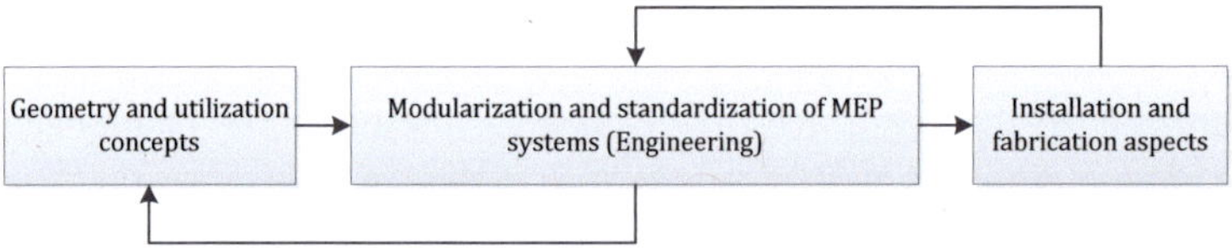

Figure 7.2: Process of implementation of M&S

Figure 7.2 shows, the key processes of implementation according to the case studies, depending on integrating teams of geometrical, utilization, engineering, installation and fabrication aspects.

The goal is to develop a model as a guideline for the implementation of M&S of MEP systems. The case studies were used to discover a design methodology and challenges in implementation, which have been further analyzed to define elements of the implementation of M&S. The implementation of M&S can be described as an interconnected process. Defining M&S as an output is the first task to be pursued. The requirements for such an implementation are as follows: (1) all participants and stakeholders should be defined and must be ready to cooperate with each other to achieve the defined outputs; and (2) achieving the same understanding of the goal of M&S, because there are variety of perspectives about the meaning of M&S. Understanding the interdependencies among all the stakeholders is essential in the management process, whereby additional interdependencies will appear during the implementation. These interdependences cannot be precisely defined in advance because this is an evolution process that requires flexibility from the participants to adjust their design.

The design methodology depends on applying M&S through three main design levels: (1) the building structure, (2) utilization and (3) configurations and components. The first two levels represent a scaffold to achieve M&S of MEP systems. In other words, the work in these two levels aims to: (1) prepare the basis to achieve M&S of MEP systems; and 2) reduce the iterations in the design during M&S of MEP systems later, due to the interdependencies between MEP architecture from one side and geometrical and utilization aspects from the other. The third level relates to the M&S of MEP systems and it represents the detailed engineering process.

According to the case studies, the potential of M&S is either not completely captured or it includes types of waste that affect the reliability of the implementation.

During the implementation process, the aim is to achieve M&S of MEP systems through making changes to the building structure or utilizations of MEP systems. This process begins with analyzing variations between target M&S (this target is aligning structures of MEP systems to be modularized and standardized according to boundaries of spatial chunks of the utilization) and current properties of the design (where the structures of MEP systems are not modularized and standardized). Analyzing the previous variations helps to define alternatives to reduce these variations. However, these alternatives should be defined in detail and implemented to check the results and obtain feedback, as can be seen in Figure 7.3.

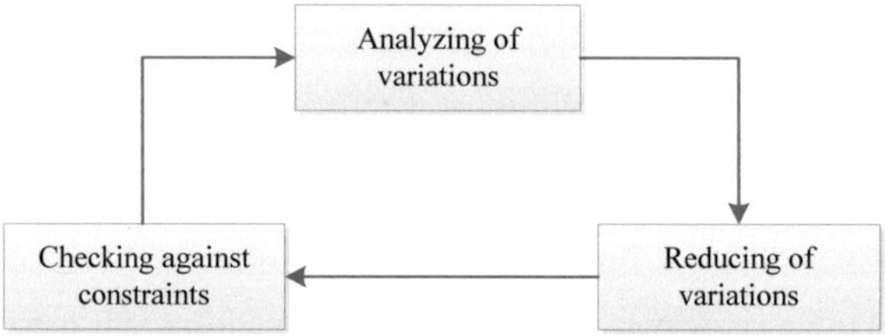

Figure 7.3: Improvement process of M&S

Making changes to increase M&S could affect other customer values like those of the user, owner or designer. Making such changes during the design aims to improve the design quality (according to the idea that M&S improve quality of the building), although it can also negatively affect the design quality if it does not consider other values and trade-offs. The following figure explains these relationships:

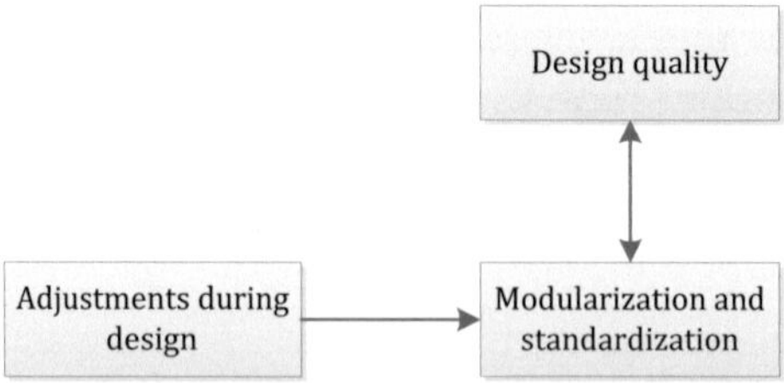

Figure 7.4: Relationship between M&S and design quality (problem solving)

Figure 7.4 highlights the following aspects: (1) M&S are made through adjustments during design; (2) M&S improve value through improving the building quality; and (3) adjustments to increase M&S could negatively affect the design quality if the interdependencies (hand-offs) are not defined and managed. The adjustments to increase M&S should be defined through the communication and development of alternatives. However, it cannot be defined in advance how M&S could affect the design quality, because this is an evolving process.

7.1.1 Workflow during the implementation of M&S of MEP systems: Making the process of M&S lean

A detailed workflow of the implementation will be explained in the following sections. The workflow is developed based upon case study 1 (modeling of a design methodology of M&S), an analysis and literature review. Another goal of describing the work flow is characterizing the processes of M&S.

7.1.1.1 Level 1: Building structure

The building structure has reference points to MEP systems (Khanzode 2011). Therefore, developing and adjusting the building structure depending on certain rules will facilitate M&S of MEP systems that will be made later during the design (detailed engineering). The development and adjustment of the building structure includes the following sub-processes:

1. Developing the building structure and defining a grid system: an initial grid system and initial positions of the geometrical elements must be developed. According to an interview, defining a proper grid system helps developing modules of MEP systems later in the design process and facilitates standardization of the modules.

2. Modularization process: The modules of the building structure are spatial chunks of the building where any geometrical element (some typical elements like shared walls, columns and facade elements) should only be contained within the boundaries of one chunk. The boundaries of the chunks are identified through the grid system. A standardized grid system has possibly identical grid fields. The size of the grid fields must be made in a way that allows for a maximum modularized and standardized geometrical model (according to case study 1). It must be mentioned that it is difficult to standardize the grid system completely, although it should be tried to increase the standardization of the grid system in a continuous improvement process, as will be explained in the next sections.

   The position of every element of the building structure must be checked, whether it is located completely in one field or on the gridline, or neither. This is the basis for a communication process in design to reposition the elements that are not completely in one field or on the gridline. The undesired position of one building element could be due to properties of some building elements or the grid system itself, etc. The task of the communication process is to identify options to improve (or increase) modularization through: (1) identifying the elements that must be changed to achieve a modularized grid system; (2) identifying the causes of a non-modularized grid system; (3) discussing the possibilities to modularize the building structure (to obtain the elements in one field or on the gridline); and (4) implementing the options of adjustments. Figure 7.5 represents the modularization process of the building structure model.

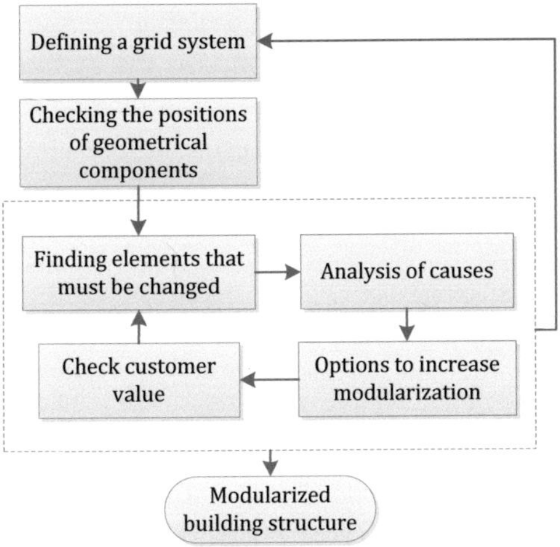

Figure 7.5: Modularization process of the building structure model

Thus, the modularization process is an iterative process. Integrating the stakeholder – who will be affected by this process – at this point in time is essential as a lean concept to ensure improving the workflow or development performance. The modularization process comprises increasing modularization of the building structure while considering the constraints of design quality.

3. Defining types of modules: a module type comprises similar modularized chunks of the building structure model. Analyzing the chunks of the building structure helps to define module types. The similarity of modularized chunks can be hardly defined; for example, the grid fields located in the corners of the building are similar because they have outside walls on two sides and thus constitute a type of field (as previously presented in case study 1).

According to the interviews, defining the similarity of modules depends on the experience of the architect. Therefore, developing criteria to define types of modules will significantly help to facilitate this process. The aim of defining of types of modules is to facilitate the standardization process.

4. Standardization processes: this is undertaken on two levels:

- Reducing or eliminating the differences within one type of modules by aligning the structure of chunks that belong to the same type of modules by making small changes in the building structure.
- Reducing the types of the modules by reducing or eliminating the differences between types of modules. This can be also achieved by making small changes to the building structure.

In the standardization process, it is essential to develop alternatives to increase the standardization, where responsible teams for the utilization, architecture, structure and MEP can participate in developing and evaluating the alternatives of standardization. Set-based design can be used in this process, although its detailed application should be checked practically. Figure 7.6 represents the iterative standardization process:

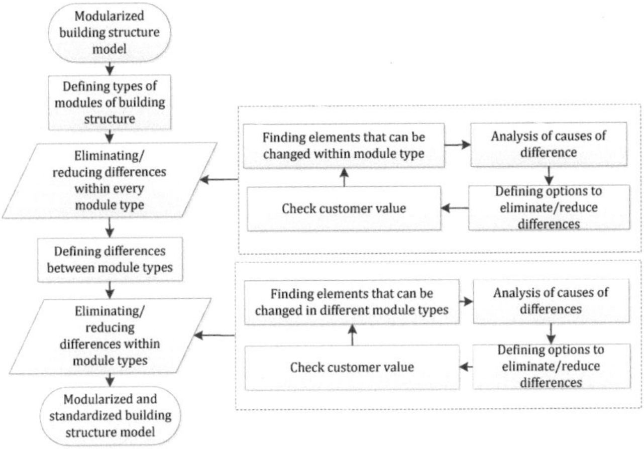

Figure 7.6: Standardization process

The question is: what differences could be reduced or eliminated? This process is an iterative one - as previously explained - and it may require developing many possible alternatives. Therefore, the use of set-based design (as previously mentioned) and pull planning (LPS) will help to analyze and implement options of M&S, while improving the performance of the implementation. The conditions of satisfaction (or evaluation criteria) among the customers can be defined as follows: eliminating or reducing the differences within and between types of modules without reducing the design quality.

To improve the efficiency of the performance, pull planning and set-based design should be implemented collaboratively through the participation of all affected teams and not by handing-off the documents between the stakeholders, which could cause a loss of information and time.

The following flowchart in Figure 7.7 summarizes the work at the level of the building structure. The iterations between the modularization and standardization processes strongly influence these processes, although they are not shown in the flowchart.

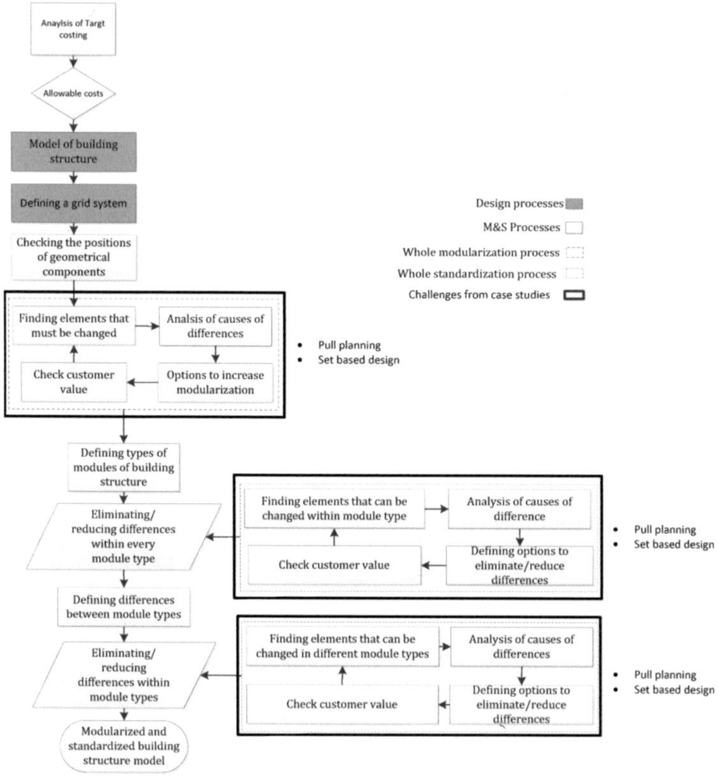

Figure 7.7: Work in Level 1- Building structure

As concluded from the case studies, the increased costs due to M&S represent a major challenge. These costs mainly result from the greater efforts required in design and the additional material costs. However, the savings achieved from

M&S are not recognized completely; for example, savings in the construction process (e.g. installation, managing and controlling on the construction site, logistic) and construction time, as well as savings in planning time and efforts in the case of changes in the layout during design. Recognizing all of these savings and integrating them in calculating the allowable costs in the target design model helps to motivate the owner to implement M&S and provides him with greater readiness to accept additional "current" costs in one category (e.g. material, planning costs in the first design phases).

7.1.1.2 Level 2: Space utilization

Developing modularized and standardized MEP systems strongly depends on the requirements of utilization and types of utilizations. The development of a hierarchy for utilizations – as explained in Figure 7.8 – helps to develop modules of utilizations. M&S of utilization can be explained as follows:

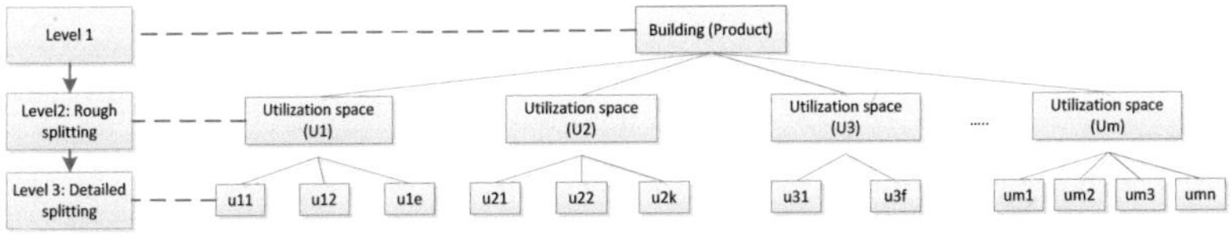

Figure 7.8: Developing units of utilization (example)

In the previous example, the rough utilization spaces are U1, U2, ...Un, which can be divided into sub-utilizations u11,.....un4. One module type contains similar sub-spaces of utilizations. Many types of modules could be defined; for example, module type 1, module type 2.

1. Modularization process:

   This process aims to align boundaries of spaces that have different (or similar) utilizations to the interfaces between fields of the grid system (Mohamad et al 2013). A utilization module is a chunk of the building with a certain space and boundaries aligned to boundaries of the grid system, and it has a type of utilization. One utilization space (like a room or part of a room) may include one or more fields of the grid system. Many types of modules could be developed.

   The modularization process includes adapting the spaces of utilization whereby their boundaries are aligned with the interfaces of the grid system, or adapting the grid system itself to be aligned with the boundaries of the utilization spaces. The modularization process includes: (1) a comparison between types of grid fields and spaces of utilization; (2) developing options to increase the alignment between types of grid fields and spaces of utilization (many options could be developed); (3) performing the options; and (4) evaluating the options according to the design quality. Design quality can be evaluated by the project teams, according to each of their perspectives and values. Using set-based design and developing trade-off curves will help to make faster decisions and improve the negotiation process to adapt suitable solutions. Figure 7.9 represents the modularization process:

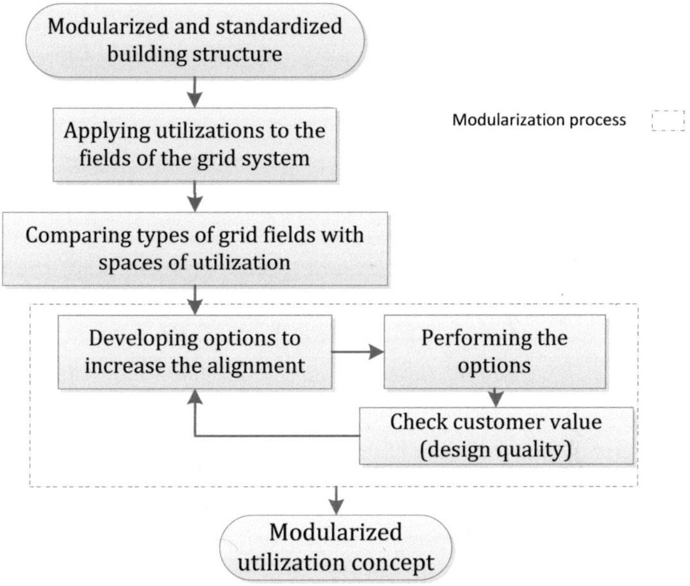

Figure 7.9: Modularization process of utilizations

2. Developing of types of modules:

   The definition of types of modules depends on the components of the modules, which are as follows: dimensions of the spaces of the modules (buildings chunks), interfaces and utilization of the spaces. The interfaces are the interfaces that affect the MEP design process. In a previous example (all corner fields of the building shall fall into the same type of space utilization), the interfaces are interfaces with a surrounding environment plus with other building spaces. Other types of interfaces should be defined by the designer.

   One type of modules includes chunks of the building that have similar components. Most likely, the process of defining types of modules significantly affects the iterations in design.

This process is a critical process because it depends on finding similar modules of utilization, and this similarity should consider all utilization factors like geometry and utilization types in the spaces. Defining this similarity will be followed with changes to obtain identical utilization spaces. How these changes affect design quality and what information is necessary to implement them should be made using pull planning and LPS to avoid losing important information.

Therefore, great attention must be paid to this process through the participation of stakeholders and the use of a DSM. However, the required data to inspect the improvement in performance and iterations caused during this process were not available within the time of this research.

3. Standardization process:

   The standardization process includes standardization within one type of module of utilization and standardization between different types of modules of utilization. Through the standardization process, many options could be developed to adapt the structure of "modules of utilizations" as changing dimensions of space or changing types of utilization of some spaces, etc. Therefore, using set-based design to generate and evaluate the design options from different perspectives is important to reduce iterations and improve performance. Accordingly, using LPS is essential to manage interfaces and commitment to achieve this process. Figure 7.10 represents the standardization process of the utilization:

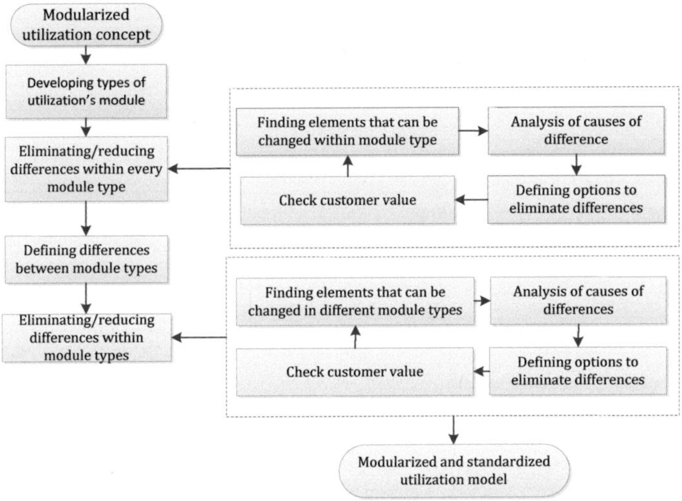

Figure 7.10: Standardization process of utilization

The following flowchart in Figure 7.11 summarizes the work at the utilization level. The iterations between the modularization and standardization processes strongly influence these processes, although they are not shown in the flowchart.

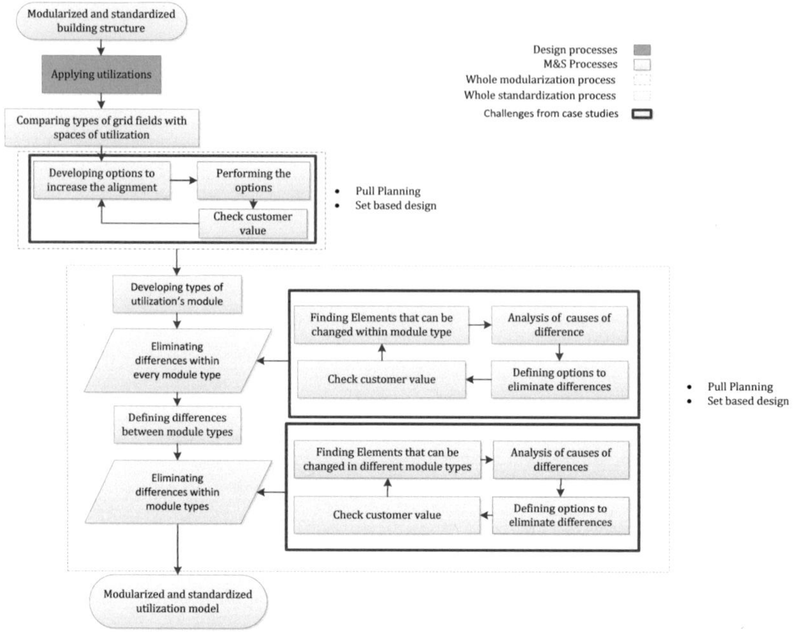

Figure 7.11: Work in level 2 - space utilization

7.1.1.3 Level 3: Configurations and components

In the modularization process, the aim is to align boundaries of systems to those of fields of the grid system (Mohamad et al. 2013). In this way, the utilizations (or functions) are mapped to the fields of the grid system.

The inputs to the modularization process are the developed modularized and standardized utilization modules, which define the strived boundaries of MEP systems. The modularization process comprises designing of MEP systems for the "modules of utilization", which includes applying design factors of MEP systems and analyzing the dependencies and the constraints within the modules. The design of the modules is an integrative process that need inputs and coordination of all MEP systems within the modules. According to case study 7, this process includes the following challenges: changing the method design should be developed, coordination of new interfaces, availability of required information at the right time during the design.

The standardization process comprises standardizing MEP systems within one utilization type of modules and between different types of modules. The standardization process of MEP systems requires analyzing the differences between the modules and developing alternatives to reduce or eliminate these differences with alignment to the design quality (different perspectives according to the different project teams) and end-customer value (flexibility and cost). These differences could be structural, dimensional or other types; therefore, an integrated approach should be used to manage the information flow to guarantee having the required information at the right time. Moreover, using set-based design to keep alternatives of standardization (according to the different perspectives of project teams) is essential to reduce iteration. According to case study 7, the process includes the following challenges: increased costs due to standardization (mostly over-design), availability of required information at the right time. Thus, pull planning and LPS can be used to manage this process. Increased costs should be aligned to the target costing.

The following flowchart in Figure 7.12 represents the work at this level. The iterations between the modularization and standardization processes strongly influence these processes, although they are not shown in the flowchart.

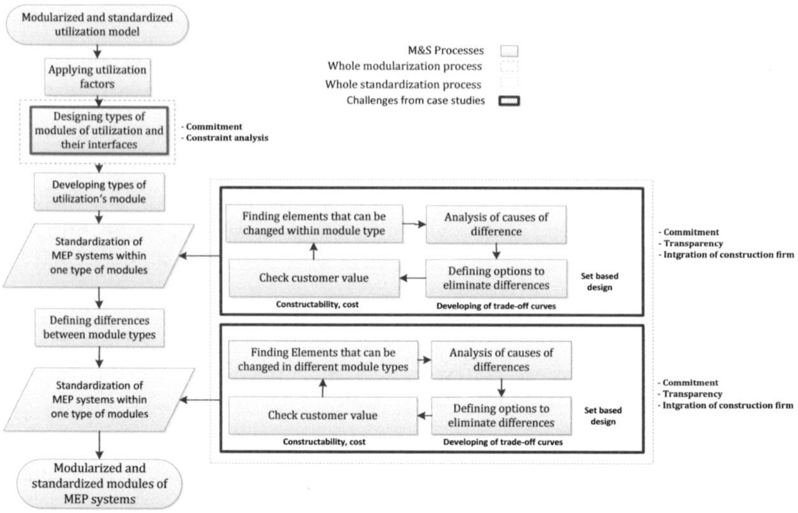

Figure 7.12: Work in Level 3 – configurations and components

The modularized and standardized structure affects the production process and supply chains on the construction site, as explained in the case study. The integration of the construction teams and installer is essential in this process to analyze the effects of positioning boundaries of MEP systems on installation, transportation and construction costs.

In the modularization and standardization processes, many criteria should be considered and integrated in the decision-making process. Trade-offs (and choosing by advantages) analysis can be used to improve the decision-making process based upon the criteria of material costs, planning costs, flexibility, operation costs and design change costs.

As a result, M&S processes are evolution processes and they can only be made through negotiations and adjustments that cannot be defined in advance.

By applying LPS, some milestones can be defined; for example, designing modules of utilizations because these milestones are strived outputs. Evaluation of the reliability of the plan is a part of LPS as a way to improve this reliability. Reliability will be discussed in the next section.

7.2 Developing two indicators to improve the reliability of the implementation of M&S

Flow is described as "the progressive achievement of tasks along the value stream so that a product proceeds from design to launch, order to delivery, and raw materials into the hands of the customer with no stoppages, scrap or backflows."(Womack 1996). The previous case studies have shown many types of waste that affect the workflow in the design and construction processes.

During construction: negative iterations (e.g. Case study of sprinkler system), waiting for decisions (e.g. Case study of sprinkler system and case study 7), resource capacity due to re-work on the construction site (e.g. Case study 3). During implementation in design: many types of waste arise according to the case study 7, re-work due to late decisions and late start (e.g. Case study 6). The reliability of implementation during the design will be discussed in the next section.

Two indicators are developed to assess the efficiency of the implementation and the reliability of the planning process during the implementation of M&S in design: (1) compatibility between modules and layout; and (2) the need to adapt interfaces within and between modules.

The two indicators are derived from the architecture of the modular design and some types of waste defined previously in the case studies. The indicators should be used to define improvement procedures by the design teams to increase the efficiency and effectiveness of the implementation in a continuous improvement process, which improves learning through implementation.

The modular architecture of a product is defined through the modules and their components from one side, as well as the interfaces between and within the modules from the other, where the process of integrating the modules is an important part to achieve the modular architecture (Baldwin and Clark 2000, case studies). Types of waste captured from the case studies define the unnecessary additional efforts needed to achieve modular architecture. The processes of defining the dimension of the modules and adjusting interfaces are defined as the processes that cause more efforts and include the most waste during implementation. Thus, making these processes more efficient will increase the reliability and improve performance. Both indicators can be explained as follows:

Compatibility between modules and layout: the goal is to achieve high compatibility between detailed modules (MEP modules) and previous modules (modules of building structure and utilizations). This indicator can be measured as a percentage part in every phase.

The incompatibility between - for example - modules of MEP and the developed building structure and utilizations is defined as a cause of waste and iterations, which lead to additional unnecessary efforts. The additional efforts are mostly caused by not considering important dependencies between different types of modules. Measuring of this indicator in the context of applying LPS will allow the participants themselves to discover the important previous dependencies and root causes for not considering them, which led to the incompatibility and then suggest methods to improve it.

Need to adapt interfaces between (or within) modules in every design level: this need is defined as a cause of additional (unnecessary) efforts to achieve the modular architecture of the whole building. The modular architecture in every phase has more interfaces than else; therefore, increased efforts to adjust interfaces is an indicator for possible increased unnecessary efforts during the implementation. It should be mentioned that some of these efforts relate to the normal evolving design process, while some of them relate to not considering important interfaces at the right time. The task of the design team is to distinguish and classify the need to adapt the interfaces. Finding root causes of increased unnecessary efforts needed to adjust the interfaces helps to define important dependencies and then improvement procedures to reduce the required efforts.

In the last case study, design project teams have defined the "insufficient awareness about important interfaces in the product structure" as a result of insufficient knowledge about the importance of these interfaces by project teams who can define these interfaces and those that need knowledge about theses interfaces. Other causes for adjusting the interfaces are different levels of detailed knowledge, which prompt the need to undertake multiple adjustments for the same interfaces.

As previously stated, the two indicators should be measured and used to define the improvement procedures as suggested in Figure 7.13:

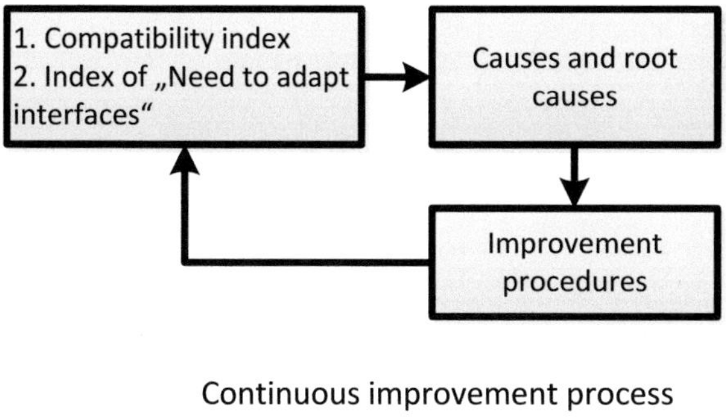

Figure 7.13: Indicators to improve performance of implementation

The definition of improvement procedures should be undertaken by project teams after defining priorities of eliminating root causes or the ability of the firm (or project teams) to eliminate them. Defining priorities is important because not all causes can be eliminated at once. This is a continuous improvement process during implementation to define the required design factors (level of design detailing) at the right time during implementation and then integrating the responsible projects' participants at the right time during the implementation. This is aligned to the challenge of defining the right detailing of design during the implementation.

Table 7.1 can be used to track the updates of the layout and interfaces and define improvement procedures from the perspective of the participants:

Table 7.1: Matrix to track required update

Adjustment	**Cause/root cause(s)**	**Participants**	**Could be avoided**	**Improvement procedure**
1				
2				
3				
4				
..				

Organizing the updates in this table aims to motivate the design teams by participation to derive improvement procedures. The following situation from the case study 7 is an example to explain this:

Adjustment: One machine needs a special interface, which is first thought about late during the design. This late thinking caused the need to adjust interfaces, which were designed before and then iterations and other adjustments for structural, architectural and utilization issues.

Cause: The user did not think before that this interface was important for the work of the designer, and the designer did not have the sufficient knowledge to know that this interface existed to consider it during the design.

Participants: The user, designer, architect and structural designer participated separately in this process, which means that the information was handed off from one participant to another.

Adjustment could be avoided: according to the design team, this adjustment could have been avoided if the participants were in the same place and if they defined the interfaces more precisely (more detailed).

Improvement procedure: the improvement procedure should answer the question of how to make important interfaces visible for the design team at the right point of time. This question should be discussed by the participation of the design teams and management to define and check the possibility of adapting defined improvement procedures.

7.3 Trade-off curves

According to the interviews, there are challenges that make the owner and development team unsure about the implementation of M&S, namely manufacturing quality and warranty claims. These challenges occurred mostly for the following reasons:

1. The existence of many interfaces due to modularization increases the probability of failure during manufacturing and installation in comparison with systems with fewer interfaces.

2. M&S of MEP systems sometimes require positioning the pipes in narrow places, where it will be needed to warranty for required installations and facilitating these installations.

In case study 3, integrating the sub-contractor helped to develop the support tools for installation, which in turn helped the designer to pursue the solution (of M&S). However, this went slowly and without strong satisfaction by the designer, which affected their performance and productivity. Analyzing the manufacturing factors such the dimensions and distances will help with integrating the sub-contractor developing installations based upon a scientific tool. Furthermore, this will also help in the negotiation process to identify what knowledge is necessary from which participants, and improve satisfaction among the participants.

Trade-off curves are a lean tool used to avoid development re-work in the later stages and to improve the design quality (Ward 2014). Manufacturing quality and warranty claims are among the factors of design quality that can be analyzed with the help of trade-off curves. The aim of using of trade curves during the implementation of M&S is to reduce uncertainty by the design teams and the owner and generate usable knowledge that improves feasibility learning, which is one of the important factors in the lean product development.

According to Ward (2014), developing trade-off curves can be started in a series of three workshops, where the first step has a focus on a certain problem, before defining causal factors and counter-measure analysis and their interdependencies. Collecting the required data to produce the trade-off curves is achieved through experimentation, analysis or simulation (Ward 2014).

The correlation between product architecture and organization structure and product quality (such as warranty claims) has been studied by Gokpinar et al. (2010), who impose the importance of analyzing the alignment between product architecture and organization structure to improve communication between design teams, which in turn improves product quality. Using trade-off curves improves communication because it requires identifying causal factors and their dependencies to ensure manufacturing quality and product performance. These factors and dependencies should be defined from different teams, who could be structural, technical or manufacturing designers. Thus, this trade-offs foster communication between design teams.

Manufacturing quality depends on design parameters and manufacturing methods. Installers or manufacturers are the people who know more about available and required tools, for which they are responsible. The cycle of creating value (Look- Ask- Model- Discuss- Act) (LAMDA) (used by Ward 2002, Ward 2014) can be used to develop and communicate the trade-offs between the teams, as suggested and explained in Figure 7.14. This process (or its results) affect the decision-making process during the implementation of M&S.

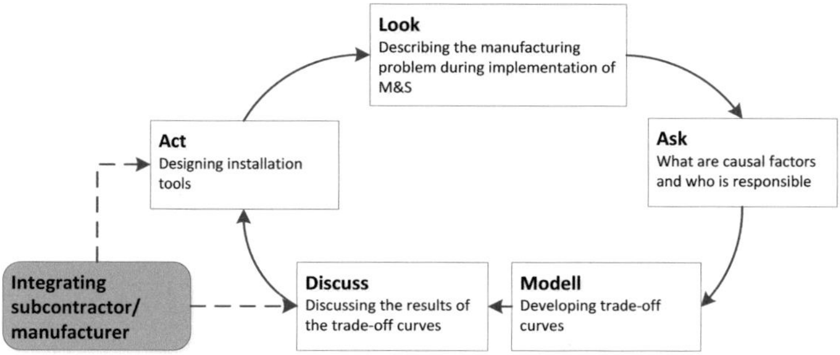

Figure 7.14: LAMDA by applying Trade-offs curves

Figure 7.15 explains benefits that can be get from developing of trade-off curves.

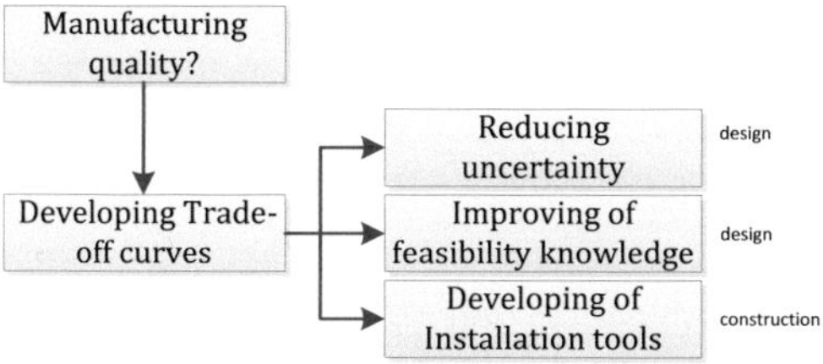

Figure 7.15: Benefits from trade-off curves

Flexibility of the organization structure is very important to develop trade-offs, because according to every situation, different project teams should be included in this process, which cannot be defined in advance. Accordingly, the organizational structure should be based upon an integrated flexible form.

7.4 The whole management system

The developed management system comprises three main elements that are important to achieve M&S of MEP systems efficiently. Every element is a process of a continuous improvement and thus every element provides feedback to improve challenges and inefficiency. Figure 7.16 represents the three elements:

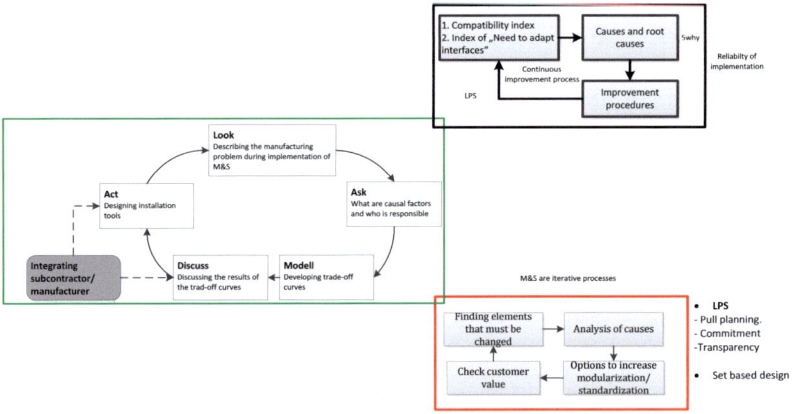

Figure 7.16: Guidelines model to manage implementation of M&S

The elements of the management system in Figure 7.16 consider the following aspects:

1. M&S are iterative processes: According to case study 1 and further analysis, M&S are iterative evolving processes that

cannot be defined in advance, and they have many interfaces with other processes (and participants). Defining these processes close in time to their execution with a focus on only the required information is essential to achieve them efficiently. Pulling required information and commitment to deliver them by the responsible participants and transparency in evaluating design solutions are elements of a successful implementation.

2. M&S are evolving processes that affect the quality of design: M&S can influence design quality in many ways; for example, manufacturability is mostly not considered at the right time and it could lead to warranty claims or adjusting design to increase M&S does not meet some design factors or guidelines. These effects on the design quality can be reduced using trade-off curves, as previously explained.

3. Two indices can be used to improve the reliability of design planning during the implementation of M&S by identifying the suitable detailed design during implementation, and motivating the participants for more cooperation.

7.5 Validation

Validation is analyzed through a case study and discussions with the design teams. The case study validation (included in case study 6) was achieved by analyzing the results of integrating construction firms to develop a standardized structural system as an element of LPS. The interviews have shown dissatisfaction by the designer about the developed system, whereby they think that the developed system will not save costs and that the construction firm will use the opportunity to increase the construction prices. This probably could have occurred because the construction firms were not transparent in introducing the costs of their system. Moreover, the late integration of the construction firms caused iteration and re-work in the design, due to handing-off the standardized architecture of the structural system from construction firms to the designer, who should check and re-design their system to be aligned with the new structural system. It can be said that the conventional management system - based upon hand-offs and the definition of tasks that are made centrally by the decision-maker (in this situation the manager and owner) - reflects the cause of the inability to identify a solution that satisfies the project participants. The designer did not understand for what benefits they must adjust their systems. In particular, a lack of participation by the designer in developing the standardized structure, as well as decisions being made by people who did not participate directly in the process itself (owner and project manager) resulted in the development of solutions that did not satisfy all participants. As a summary:

- The project manager was interested in reducing the construction time, which was achieved by standardizing the structural system by the construction firm.
- The construction firm developed the standardized structure and handed it off to the designers to adjust their systems, whereby the designers did not co-develop the standardized architecture.
- There was high risk that the construction firms would exploit this situation to quote high prices due to a lack of competition.
- Savings in costs due to standardization will not be shown in the calculation because the calculator and the calculation system do not consider savings in construction time as a part of savings in costs.
- Although the construction firm has developed an effective solution from its perspective, the designers argued that the standardized structure would increase the costs.

According to the previous results, it can be argued that the integration of the construction firms during the design (as part of applying LPS) was not sufficient. It can be claimed that the integration of construction firms to achieve standardized (target) architecture was not lean, because it was not based upon collaboration between project teams. This confirms the impacts of a traditional management system on the satisfaction of project partners and then on their transparency and their willingness to introduce usable knowledge, when target architectures for the building systems are strived.

Using set-based design was argued by the designer to take more time in design. This is aligned to findings in literature, whereby the designer – who did not use set-based design – was unable to validate its positive impacts on iterations and design time.

In terms of developing two indicators to measure the reliability of the implementation, the design team agreed that reliability is not sufficient and should be improved, and that indicators can be used to define design factors at the right time and then integrating the responsible participants at the right time; however, these indicators are difficult to measure, while integrating of more participants in the design discussions (e.g. workshops) will increase the cost of planning, and the owner will not be willing to pay for these costs. Therefore, future research should focus on measuring the indicators in a real project to show the possibility of measurement and the improvement that could be achieved. The availability of suitable conditions for this should be studied and ensure achieving success.

Table 7.2 summarize aspects from the case studies and elements of the developed guideline of the management system to manage the potential of M&S of MEP systems. The implementation of the elements presented in the first and third columns is suggested to improve its performance, which should be validated in a practical case study.

Table 7.2: Aspects from the case studies and element of the developed guideline of the management system

Successful aspects from the case studies	Aspects (causes) for improvement from the case studies	Element of the developed management system
• Using a structured design methodology (case study 1). • Early implementation of M&S in the design reduces iterations later during the implementation and leads to early adjusting the geometry of the building as a prerequisite for M&S of MEP (case study 2). • Integration of construction firm as part of the last planner system (case study 6-2). • Integration of user (case study 6-1).	• Sequential implementation in design can reduce potentials of M&S of MEP systems (case study 6-1). • Integration of installer during implementation (case study 4). • An incentive system is required to improve cooperation and create satisfaction among the participants during the implementation (case study 4). • Costs should drive the implementation in design to achieve customer value (case study 4). • Integrating construction teams during the implementation is required to evaluate the criteria of the decision-making related to construction process transparently (case study 3). • The alignment of customer value is required (case studies 3, 6-2). • Transparency and willing to cooperate early in design to reduce iterations (case study 2). • Thinking of some project participants (such as user in case study (6-1)) must be oriented to achieve global optimization and not the local ones (case study 6-1).	1. Managing the iterative processes of M&S through: • LPS to achieve: - Commitment - Transparency - Pull planning - Real time alignment of customer values. • Set-based design to achieve: - Reduce iterations. • Using trade-off curves to achieve: - Reducing uncertainty during implementation. - Improving feasibility knowledge during implementation. • BIM to achieve: Integrated implementation on aspects of utilization, architecture and detailed engineering of the different systems. • Framework conditions to assure a proper atmosphere to implement the previous elements.

	• M&S take a very long time due to using of point based design, which caused iterations (case study 7) . • Transparency, controlling and commitment to improve product-ion during implementation are required (case studies 1, 7).	
Integration of installer during imple ment tation in design to: 4. convincing the designer of the constructability of the system, 5. developing installation supports to be delivered on time to the construction site, and to facilitate construction process (case study 4).	• Worries about warranty claims due to the modularized and standardized structure of MEP systems (case study 4, interviews). • Risk of performance of installation (case study 4, interviews).	2. Improving quality of design during implementation through: • Using trade-off curves to achieve: - Developing installation methods and tools. - Improving communication during implementation. • Integration of construction firms during the implementation to achieve: - Understanding the system of M&S - Evaluating of criteria related to construction aspects transparently (e.g. cost factors).
	• Defining the right detailing of design at the right time during implementation is a challenge, which causes iterations (case study 7, Experts).	3. Improving the reliability of the implementation through: • Using reliability indicators to define required design factors at the right time in a continuous improvement process.

	• Types of waste during implementation are related to the defined boundaries of the modules and considering of interfaces-related design factors at the proper time (case study 7). • Measuring reliability of planning in relation to the modularized and standardized structure can help participants to define required design factors and root causes of unsuccessful definition of these factors through a continuous improvement process (case study 7).	The indicators are: - Compatibility between modules and layout - Need to adapt interfaces between (or within) modules. • LPS to define: - Required design factors (for future projects) - Root causes of unsuccessful definition of these factors (for a continuous improvement process in the same project to improve performance).
Framework conditions		
	• Integrated project development will reduce types of waste (especially, hand-offs and scatter) (case study 7). • Not aligning all customer values will cause iterations and possible missed opportunities to maximize customer value (case studies 3, 5, 6-1, 6-2, 7). • Hand-offs lead to dissatisfaction and not understanding the developed systems (case study 6-2). • Early implementation should be ensured to reduce iterations (case study 6-2).	4. IPD

	• Transparency is required by calculating construction costs and avoiding increased prices by using other types of contracts such as IPD (case studies 5-2, 6-2). • The calculation of costs should be based upon precise knowledge (case study 6-2). • M&S of the structures of MEP systems were not made due to the perspective that: (1) this is the task of the construction firm, and (2) this will cause more efforts in design which is not considered by HOAI (case study 6-1). • Tend to make local optimizations during construction (case study 5) • Integration of construction firms (construction partners) during the implementation (case studies 5, 6-1). • An incentive system is required to achieve cooperation and create satisfaction among the participants during the implementation (case study 4). • Willing to cooperate early in design (case study 2).	
	• The increased costs can be better analyzed by integrating lifecycle costs and benefits (using of TVD) (case studies 4, 7). • Cost should drive the implementation in design to achieve customer value (case studies 4, 7, 6-1).	5. TVD

From the previous analysis, the impacts of making M&S can result in project cost risk and performance risk. The risk of project cost results from the added costs due to M&S due to the need for more material and more efforts in design. The risk of performance results from the warranty issues during construction and operation. Figure 7.17 represents the effects of the suggested management system on reducing the risks resulting from the implementation of M&S:

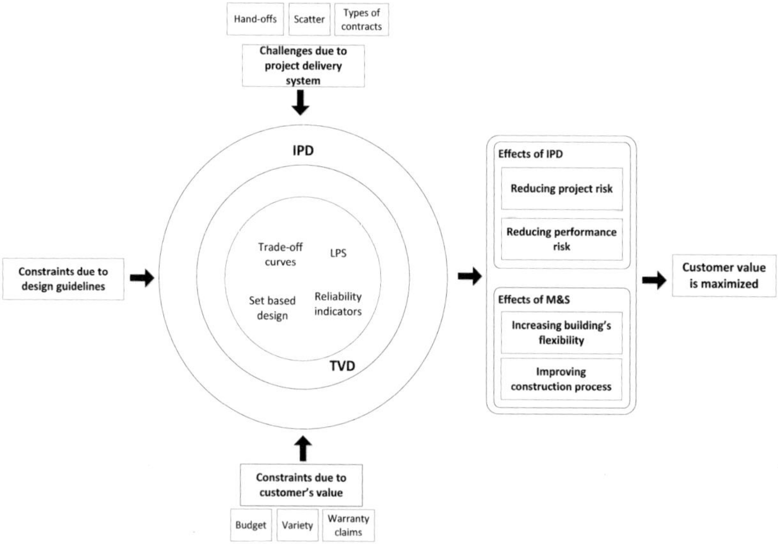

Figure 7.17: Effects of the suggested management system on reducing the risks resulting from the implementation of M&S

8 Conclusions

The research has analyzed the modularization and standardization of mechanical, electrical and plumbing systems as a way to shield the design from external unforeseen design changes, increase the flexibility of the building and enable mass production through reducing the variety of components. The conclusion of the research can be summarized in the following points: (1) describing the adaptability of implementing of M&S on MEP systems; (2) exploring the challenges to implement M&S according to the design method used; (3) linking the challenges in the traditional management system used to manage construction projects; (4) exploring the types of waste during the implementation; (5) characterizing M&S processes; (6) developing two indicators to improve the reliability of the implementation; and (7) describing a continuous improvement process during the implementation, as follows:

1. Analyzing and describing the adaptability of the implementing of M&S on MEP systems. First, analysis of the literature about the utilization and benefits of using M&S in the product development has shown that flexibility against design changes and the improvement of production systems can be achieved by applying M&S. Achieving these benefits by improving architectures of MEP systems was the motivation to inspect the possibility of this adaptability. This adaptability was inspected in a case study, which showed that this is possible. The case study with further analysis linked to literature introduced a detailed description of modularization and standardization, which can be used as a reference for future projects.

   However, the case study showed possible conflicts and challenges, which inspired further analysis of the design method used and the challenges. One of the important prerequisites of adaptability is the early implementation during design, which is shown through a case study explaining the importance of this early implementation to reduce re-work and iterations later.

Moreover, the research showed similarities between the building industry and other industries in applying M&S on MEP systems, where the modular structure is achieved - as in the other industries - through mapping functions to modules of the product, which improves flexibility and the construction process.

2. Exploring the challenges to implement M&S according to the design method used (which was analyzed in a previous case study): the challenges can be classified into design challenges and construction challenges. The design challenges arise during the design and are caused due to applying the design methodology itself and the current management system of design used. The design methodology of M&S depends on modularizing and standardizing building structures, the utilization and architecture of MEP systems, which requires analyzing and finding design solutions for many more interfaces within the product (building), in comparison to the normal situation (where no M&S is made). This means more interdependencies, which are reflected in organizational requirements. This change in the organizational structure and product structure should be managed by using lean tools and new types of contracts to achieve high-value products with minimum costs.

   Construction challenges arise because the design process did not consider the values of the construction teams and it does not consider the importance of the alignments with them. This can lead later to changing the properties of the components that are important for the modularized and standardized structure due to two causes: cost factors and warranty claims. Other types of challenges relate to the reluctance to change the traditional way of design and construction. However, it can be claimed that this reluctance also relates to the separated responsibilities, whereby not all teams participate in developing the modules, which leads them to sometimes refuse proposed solutions from cost or design perspectives, so that they do not struggle to find proper design solutions.

It could be concluded that design challenges include the insufficiency of communication occurring due to insufficient planning time, an unwillingness to change or unavailability of required information due to the absence or non-commissioning (at the right time) of required participants. During construction, changes are made by the construction firms due to aspects of decision-making and costs. However, it can be argued that managing the design process properly and integrating construction teams and the availability of a proper atmosphere to cooperate (types of contracts) can avoid challenges during the design and construction.

The decision-making process during the implementation is also an important issue if needed to increase M&S, whereby the drivers of the decision are the advantages. The barriers emerging from the disadvantages are mostly the costs, the willingness of the team to implement forward and production control during implementation. Here, the participants in this decision and how they evaluate the advantages (it should be in terms of production process and flexibility against design changes) and disadvantages (it should be in terms of additional material costs and more efforts in design) play an important role. It can be concluded here that the criteria for the decision-making processes should be based upon a deep analysis of end-customer needs and a transparent evaluation of production factors. A very important issue that could be noted from the different projects is that the additional costs emerging due to M&S could not be identified by the team of the design phase in terms of numbers or percentages.

A further possible challenge during the implementation of M&S is the standardization of "look", which is linked to variety as a value for the end-customer. Variety reflects value for the end-customer if he necessarily wants variety that causes variety in the architecture of the MEP systems. However, the variety in "look" does not necessarily produce variety in the architecture of the MEP systems. Accordingly, this is a task of the designer to explain this to the end-customer.

The general trade-off problem can be defined between increasing the flexibility and stability of workflow of MEPs' production process on the construction site and mass production from the one side and increasing efforts in design and additional material costs on the other. The negotiation process during the implementation can be explained in Figure 8.1. However, the weighing of these factors for decision-making during the implementation of M&S depends on the building type, end-customer value and transparency of project teams. Two types of buildings (offices and industrial building) studied in two case studies have a need for flexibility in the design. However, if there is little or no need for flexibility, the other advantages – in achieving mass production and a faster construction process - and disadvantages will be weighed differently.

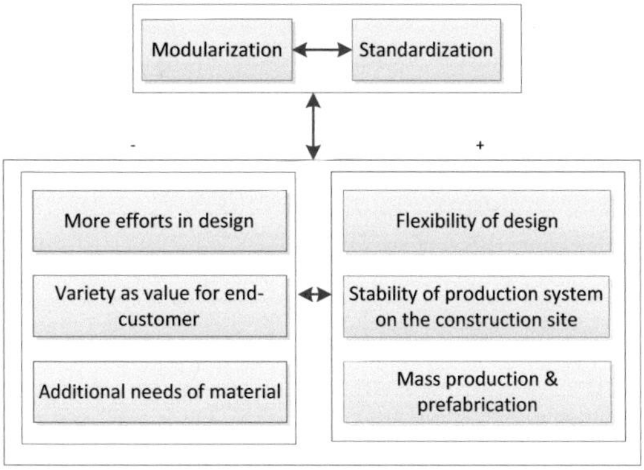

Figure 8.1: General trade-off problem during implementation

3. Linking the challenges to the traditional management system used to manage construction projects: the core causes for the occurrence of many types of challenges are the traditional management system to manage the design and construction process and the interfaces between these phases. The current management systems can be described as follows: separation of design and construction, immature decision-making in design and construction, which do not consider constructability, hand-offs and scatter. Moreover, M&S need more efforts in design; however, the paying system used does not consider the efforts required in the design but mostly depends on construction costs. The research implies that striving for M&S of MEP systems as a way to improve value is not sufficient. The contextual aspect of the project must be deeply understood and aligned with processes of M&S. It is explained that the implementation of M&S of MEP systems is interrelated with the organizational aspects that should be adapted through the integration of design and construction and use of new types of contracts to increase cooperation between the teams. The project team should have the ability to make changes (to achieve M&S) in the structure of building systems flexibly and quickly.

4. Exploring types of waste during the implementation: types of waste explored through the case studies can be described using two types of waste presented by Ward (2014): hand-offs, and scatter, such as late decisions and late begin, unshared information, late assignments of specialists, changes of priorities and decision-maker and unused employee creativity.The different types of waste led to long cycles of the learning process, which could extend from construction to the design. Exploration of the types of the wastes was used to define linkages with the modularized and standardized architecture to find causes and root causes of wastes as a way to derive improvements for the implementation during the design.

5. Characterizing the M&S processes:The aim of characterizing the M&S processes is to find a proper way to manage them according to their characteristic based upon the idea that they could be the cause for the increasing challenges and types of wastes. These characteristics were described through analyzing the modeled design methodology. As a result, M&S processes are iterative, evolving processes and many participants could be affected through the implementation, such as users, designers, manufacturers and installers. These features are mostly the reasons why many types of wastes arise through and after the implementation in design. LPS, set-based design and integrated form of contracts are suggested as lean tools to manage these processes.

6. It can be claimed that M&S improves design quality, although considering manufacturability as a part of the design quality must be ensured. This can be achieved by using trade-off-curves as a lean tool, which also helps in negotiations during development and in situations of warranty claims.

7. Applying M&S affects decision-making process in the design and then adds considerable complexity to collaboration patterns

during the design. The implementation of M&S can result in different design solutions for building systems. Therefore, the important question is what type will achieve the value of the end-customer. Decisions that include trade-offs should be made during the implementation. The trade-offs should consider the criteria of flexibility, the construction process, the standardization of components and design quality.

8. Development of two indicators to improve the reliability of the implementation:

   Based upon characteristics of modular architecture and types of waste captured in the final case study, two indicators were developed to increase the reliability of planning during the implementation in design as a way to improve the performance of the implementation. The two indicators are:

   - Compatibility between modules and layout in every design level
   - Need to adapt interfaces between modules in every design level
   - The two indicators show the linkage between the types of wastes and the architecture of the product. These linkages can be used to derive feedback about improving detailing of the design to improve the performance of the implementation.

9. Describing a continuous improvement process during the implementation. The continuous improvement is an important part of any lean model, described in Figure 7.16. The improvement process comprises measuring and evaluating the current situation (performance through measuring PPC plus the two suggested indicators), before analyzing the causes and root causes (using 5 Whys) to provide feedback to the project team to define improvement procedures.

10. Defining basics (or factors) to develop a cost model of integrating M&S in design. These factors are important to calculate the costs of applying M&S in the design. They are: flexibility, construction costs, needed time of planning due to more efforts needed in the design and the type of efforts needed by the designer.

Table 8.1 summarizes the case studies conducted during this research:

Table 8.1: Summery of case studies

Case study	Goal	Type of waste/challenge
1	Modeling of design methodology to achieve M&S of MEP.	• Modules are defined from only one perspective (designer perspective) and then handed off to the project participants, which causes iterations. • Methods to capture and organize the required information (of utilizations and detailed engineering) according to fields of the grid system (modules) are necessary. • Alignment of customer values during the implementation in design (Who are the customers? How they are affected?).
2	Point of time to begin with modularization and standardization.	• Understanding and willingness for implementation. • Having the stakeholder in this early phase.
3	Benefits of implementtation in reducing the variety of MEPs' components.	• Evaluation of the benefits is made only from the perspective of the designer. • More quantity of material is required. • Interdependencies with the ability of construction are not dealt with at the proper time in design.
4	Need for knowledge of installation during the implementation in design.	• The additional costs could not be identified in numbers or percentages. • Waste during design: unwillingness to implement, which caused unused employee creativity (people feel unmotivated). • Waste during design: defects and corrections during construction due to insufficient understanding.

5-1	Results of application of the modeled methodology (in case study 1) on the construction site- heating system.	• Savings in construction time and costs could not be realized. • Sub-contractor and construction manager have other interests. • Tend to make local optimization • Waste: post-permit design changes due to cost or constructability.
5-2	Results of application of the modeled methodology (in case study 1) on the construction site - sprinkler system.	• Different perspectives by different project participants hinder modularization and standardization. • Local optimization • Waste: - Post-permit design changes due to cost, constructability or coordination issues. - Waiting for the new decisions. - Processing: construction firms make new processing of the components.
6-1, 6-2	Design process: challenges and types of waste.	• Different interests of design teams lead to difficulties in implementation. • Integration of construction firms creates atmosphere of dissatisfaction among other teams. • Calculation system. • Payment system for the designer hinder the more cooperation required. • Waste: late decisions and late begin, iterations, waiting for decisions, and resource capacity.
7	Design process: challenges and types of waste.	• Sequential management leads to iterations. • High development costs. • Using a proper detailing design is essential. • Waste: - Modules are developed and then handed off to the other teams. - Late assignment of specialists leads to iteration. - Changes in priorities and decision-maker lead to inconsistency in negotiations and adjustments.

The studied application of M&S can be described as the design for flexibility and design for takt. However, the case studies and analysis have shown that this application should be linked to assembly issues, whereby manufacturability and warranty should be considered early during the implementation in the design phase as factors of the trade-offs.

The results of the research reveal that improving implementation can be achieved through three main elements:

1. Improving the iterative processes by using LPS and set-based design to achieve the requirements of:
   - Need to commitment and transparency.
   - Managing many more interfaces.
   - Integrate aspects of the downstreams during the implementation in the design (aspects of construction).
   - Reducing iterations.

2. Improving value generation and the quality of decision-making through using trade-off curves to consider manufacturability and warranty issues; and (3) continuous improvement process by using two indicators to improve the performance of the implementation.

Complementary to these elements, improving the implementation requires:

- A value-oriented calculation system
- The use of an integrated form of contracts
- Early implementation in design
- Integration of construction teams during detailed design

Similarly to the previous guideline, the key enablers of innovative product development defined by Bozdogan et al. (1998) can be adapted to the development process of M&S, as concluded in this research. These enablers are: long-term commitment to suppliers; co-location; joint responsibility for design and configuration control; seamless information flow; and retaining flexibility in the definition of system configuration. Important contributing factors include: supplier-capability-enhancement of investments; target costing; and incentive mechanisms.

As derived from this research, the reasons for improving the implementation are as follows:

- The product is not competitive;
- Lack of manufacturability; and
- Acceleration of the development process, which depends on the extensive use of suppliers as expert developers, which is not assured.

The developed guidelines of implementation can be seen as a standard process to manage the implementation of M&S of MEP systems and it describes objectives and boundary conditions, but not a certain solution. Improving the development performance (workflow view) and keeping the design quality (value view) can be achieved through using the suggested lean tools. The principles of LPS refer to a proper management tool during the implementation of M&S. As an aspect of applying LPS, in the look ahead phase of LPS - more precisely in the activity definition model - M&S should be part of the design criteria.

The scientific contribution of the research work is to provide evidence of the need to apply lean thinking in design to achieve modularization and standardization as a method to increase product value, while reducing waste and improving workflow during the implementation. Another scientific contribution is analyzing the impacts of the current management system on achieving systems architecture that improves product quality, whereby the efficiency of the management system can be measured: in this situation, in terms of its ability to achieve target systems architecture efficiently.

9 Future research

Based upon this work, the following ideas for future research are recommended:

1. Implementation of the inspected design methodology in an integrated project development to inspect the extent to which the project team can achieve M&S of MEP systems, as well as what challenges might occur.

2. Using set-based design during the implementation to inspect how this tool can affect iterations and decision-making process during the implementation.

3. Analysis of trade-off curves with cooperation with the project team in proper points in time and receiving feedback for the decision-making process in the design.

4. Quantitative analysis of the effects of the implementation of M&S on product variety and end-customer value in a certain case study.

5. Analysis of applying LPS on the planning process in real projects, where M&S will be implemented.

6. Analyzing what organizational changes such as types of contracts are required to implement M&S of MEP systems through empirical studies.

7. Inspecting the potential of improving the construction process in terms of improving the learning process, reducing installation time, improving pre-fabrication possibilities and the impacts on material costs using quantitative case studies.

8. The evaluation of the costs should not be based upon average values, but should also consider other factors such as adoption possibilities and fast construction process. Future research is necessary to develop indicators to evaluate the costs, as well as where these indicators can be used during design to improve the development process of M&S.

9. Analyzing the adaptability of M&S on different types of building

References

Akao, Y., ed. (1990). "Quality Function Deployment." Productivity Press, Cambridge MA.

Albers, A., Burkardt, N., Meboldt, M., and Saak, M. (2005). "SPALTEN problem solving methodology in the product development." ICED 05: 15th International Conference on Engineering Design: Engineering Design and the Global Economy, Engineers Australia, 3513.

Albers, A. and Meboldt, M. (2007). "SPALTEN Matrix–product development process on the basis of systems engineering and systematic problem solving." The future of product development, Springer, 43-52.

Austin, S., Baldwin, A., Li, B. and Waskett, P. (1998). "Development of the ADePT Methodology." : An Interim Report on the Link IDAC 100 Project. Department of Civil and Building Engineering, Loughborough University.

Baiden, B. K., Price, A. D. and Dainty, A. R. (2006). "The extent of team integration within construction projects." International Journal of Project Management, 24(1), 13–23.

Baldwin, C. Y. and Clark, K. B. (1997), "Managing in an age of modularity". Harvard Business Review, 75, 84–93.

Baldwin, C. Y. and Clark, K. B. (2000). Design rules: The power of modularity. MIT press.

Baldwin, C. Y. and Clark, K. B. (2003). "Managing in an age of modularity." Managing in the Modular Age: Architectures, Networks and Organizations, 149.

Baldwin, C. Y. and Clark, K. B. (2006). Modularity in the design of complex engineering systems. Springer.

Ballard, Glenn (2000). "Lean Project Delivery System." White Paper #8, Lean Construction Institute, May 1, 6 pp.

Ballard, G. (1999). "Can pull techniques be used in design management?" CIB REPORT, 149–160.

Ballard, G. (2000a). "Managing work flow on design projects." Atlanta, 19, 20.

Ballard, G. (2000b). "Positive vs negative iteration in design." Proceedings Eighth Annual Conference of the International Group for Lean Construction, IGLC-6, Brighton, UK, 17–19.

Ballard, G. (2002). "Managing work flow on design projects: a case study." Engineering Construction and Architectural Management, 9(3), 284–291.

Ballard, G. (2008). "The lean project delivery system: An update." Lean Construction Journal, 2008, 1–19.

Ballard, G. and Arbulu, R. (2004). "Making prefabrication lean." Proc. 12 th Ann. Conf. of the Int'l. Group for Lean Constr, 3–5.

Ballard, G., Hammond, J. and Nickerson, R. (2009). "Production control principles." Proceedings of the 17th annual conference of the International Group for Lean Construction, 489–500.

Ballard, G. and Howell, G. (2003). "Lean project management." Building Research & Information, 31(2), 119–133.

Ballard, H. G. (2000c). "The last planner system of production control." The University of Birmingham.

Ballard, G. (2006). "Rethinking Project Definition in terms of Target Costing". Proceedings of the 14th annual conference of the International Group of Lean Construction. Chile

Bertelsen, S. (2005) "Modularization –A Third Approach to Making Construction Lean?" Proceedings of the 13 th Annual Conference of the International Group for Lean Construction , Sydney, Australia.

Bertelsen, S. and Emmitt, S. (2005). "The client as a complex system." 13th International Group for Lean Construction Conference: Proceedings, International Group on Lean Construction, 73.

Björnfot, A. and Stehn, L. (2007). "Value delivery through product offers: a lean leap in multi-storey timber housing construction." Lean Construction Journal, 3(1), 33–45.

Blismas, N. G., Pendlebury, M., Gibb, A. and Pasquire, C. (2005). "Constraints to the use of off-site production on construction projects." Architectural Engineering and Design Management, 1(3), 153–162.

Blyth, A. and Worthington, J. (2010). Managing the brief for better design. Routledge.

Bock, T. and Linner, T. (n.d.). "ENHANCED INDUSTRIALIZED CUSTOMIZATION PERFORMANCE BY EMBEDDED MICROSYSTEMS."

Börjesson, F. (2012). "Approaches to Modularity in Product Architecture." Stockholm Royal Institute of Technology.

Bozdogan, K., Deyst, J., Hoult, D. and Lucas, M. (1998). "Architectural innovation in product development through early supplier integration." R&D Management, 28(3), 163–173.

Brand, S. (1995). How buildings learn: What happens after they're built. Penguin.

Brodetskaia, I., Sacks, R. and Shapira, A. (2011). "A workflow model for systems and interior finishing works in building construction." Construction Management and Economics, 29(12), 1209–1227.

Brodetskaia, I., Sacks, R. and Shapira, A. (2012). "Stabilizing production flow of interior and finishing works with reentrant flow in building construction." Journal of Construction Engineering and Management, 139(6), 665–674.

Christoferson, A. K. (2003). " value contrl in the building process." presentation at the LCI congress, Blacksburg, Virginia.

Cooper, R. G. (1994). "Perspective third-generation new product processes." Journal of Product Innovation Management, 11(1), 3–14.

Court, P. F., Pasquire, C. L. and Gibb, A. G. F. (2009a). "A lean and agile construction system as a set of countermeasures to improve health, safety and productivity in mechanical and electrical construction."

Court, P. F., Pasquire, C. L., Gibb, G. F. and Bower, D. (2009b). "Modular assembly with postponement to improve health, safety and productivity in construction." Practice Periodical on Structural Design and Construction, 14(2), 81–89.

Digitales Bauen. (2008). Modulares Bauen - Ein Kostenvergleich. (available at: www.digitales-bauen.de/pdf/Kostenvergleich_ModularesBauen_090128.pdf (last accessed 2013-06-06).

Dombrowski, U., Schmidt, S. and Schmidtchen, K. (2014). "Analysis and integration of Design for X approaches in lean design as basis for a lifecycle optimized product design." Procedia CIRP, 15, 385–390.

Du, X., Jiao, J. and Tseng, M. M. (2006). "Understanding customer satisfaction in product customization." The International Journal of Advanced Manufacturing Technology, 31(3-4), 396–406.

Egan, J., (1998). "Rethinking Construction." The Egan Report, London, Department of the Environment, Transport and the Regions.

El. Reifi, M. H. and Emmitt, S. (2013). "Perceptions of lean design management." Architectural Engineering and Design Management, 9(3), 195–208.

Emmitt, S. and Gorse, C. (2003). " Construction Communication." Blackwell Publishing, Oxford.

Emmitt, S., Sander, D. and Christoffersen, A. K. (2004). "Implementing value through lean design management." Proceedings of the 12th International Conference, 361–374.

Emmitt, S., Sander, D. and Christoffersen, A. K. (2005). "The value universe: defining a value based approach to lean construction." 13th International Group for Lean Construction Conference: Proceedings, International Group on Lean Construction, 57.

Emmitt, S. (2006). "Researching a value-based approach to design and construction". *In*: Aouad, G., Kagioglu, M., Harris, K., de Ridder, H. A. J., Vrijhoef, R. and van den Broek, C., eds. *Proceedings of 3rd International Salford Centre for Research and Innovation (SCRI) Symposium, 2006.* Salford, U. K.: University of Salford, pp. 65-77.

Eppinger, S. D. and Browning, T. R. (2012). Design Structure Matrix Methods and Applications. Mit Pr, Cambridge, Mass.

Ericsson, A. and Erixon, G. (1999). Controlling design variants: modular product platforms. Society of Manufacturing Engineers.

Erixon, G; (1996); "Modular function deployment (MFD)., support for good product structure creation"; In: Proceedings of the 2nd WDK workshop on Product Structuring, pp. 181-196; Tichem, M; Storm, T.; Andreasen, M.M.; MacCallum, K.J. (ed.); Delft; ISBN 90-370-0148-3.

Erixon, G., (1998). "Modular Function Deployment - A Method for Product Modularisation." Dept. of Manufacturing Systems, Royal Institute of Technology, Stockholm.

FABEL-Report No. 13 Druck. (1993). "Similarity Concepts and Retrieval Methods." Gesellschaft für Mathematik und Datenverarbeitung mbH (GMD), Sankt Augustin - Voß.

Faniran, O. O., Love, P. E. D., Treloar, G. and Anumba, C. J. (2001). "Methodological issues in design-construction integration." Logistics Information Management, 14(5/6), 421–428.

Galbraith, J. R. (1973). Designing complex organizations. Addison-Wesley Longman Publishing Co., Inc.

Gershenson, J. K. and Prasad, G. J. (1997). "Modularity in product design for manufacturability." International Journal of Agile Manufacturing, 1(1), 99–110.

Gershenson, J. K., Prasad, G. J. and Zhang, Y. (2004a). "Product modularity: measures and design methods." Journal of Engineering Design, 15(1), 33–51.

Gershenson, J. K., Prasad, G. J. and Zhang, Y. (2004b). "Product modularity: measures and design methods." Journal of Engineering Design, 15(1), 33–51.

Gidey, E., Jilcha, K., Beshah, B. and Kitaw, D. (2014). "The Plan-Do-Check-Act Cycle of Value Addition." Industrial Engineering & Management.

Gokpinar, B., Hopp, W. J. and Iravani, S. M. (2010). "The impact of misalignment of organizational structure and product architecture on quality in complex product development." Management Science, 56(3), 468–484.

Greden, L. V. (2005). "Flexibility in building design: a real options approach and valuation methodology to address risk." Massachusetts Institute of Technology.

Haller, F. (1985). "ARMILLA-ein Installationsmodell." Institut für Industrielle Bauproduktion der Universität Karlsruhe. Germany, 1985

Hamzeh, F. R., Ballard, G. and Tommelein, I. D. (2009). "Is the Last Planner System applicable to design?—A case study." Proc., 17th Annual Conf. of the Int. Group for Lean Construction (IGLC-17).

Hansen, G. K. and Olsson, N. O. (2011). "Layered project–layered process: Lean thinking and flexible solutions." Architectural Engineering and Design Management, 7(2), 70–84.

Hopp, W. and Spearman, M. (1996). "Factory Physics: Foundations of Manufacturing Management." Irwin/McGraw-Hill, Boston, 668 pp

Hickethier, G., Tommelein, I. D. and Lostuvali, B. (2013). "Social network analysis of information flow in an IPD-project design organization." 21st International Group for Lean Construction Conference Proceedings (IGLC-21), 319–328.

Holt, R. and Barnes, C. (2010). "Towards an integrated approach to 'Design for X': an agenda for decision-based DFX research." Research in Engineering Design, 21(2), 123–136.

Hölttä-Otto, K. and others. (2005). Modular product platform design. Helsinki University of Technology.

Israelsson, N. and Hansson, B. (2009). "Factors influencing flexibility in buildings." Structural Survey, 27(2), 138–147.

Jensen, P., Hamon, E. and Olofsson, T. (2009). "Product development through lean design and modularization principles." Submitted to IGLC, 17, 2009.

Jensen, P., Olofsson, T. and Johnsson, H. (2012). "Configuration through the parameterization of building components." Automation in Construction, 23, 1–8.

Jensen, P. and Rönneblad, A. (2010). "Configuration through the parameterization of building components." Proceedings of the CIB W78 2010: 27th International Conference –Cairo, Egypt, 16-18 November.

Jiao, J. and Tseng, M. M. (2004). "Customizability analysis in design for mass customization." Computer-Aided Design, 36(8), 745–757.

Jørgensen, B. and Emmitt, S. (2009). "Investigating the integration of design and construction from a 'lean' perspective." Construction Innovation: Information, Process, Management, 9(2), 225–240.

Kano N., Seraku N., Takahashi F., Tsuji S. (1984.) "Attractive quality and must-be quality. " Hinshitsu Kanri 14(2),147–156.

Khanzode, A. R. (2011). An integrated virtual design and construction and lean (IVL) method for coordination of mechanical, electrical and plumbing (MEP) systems. Dissertaion

Koskela, L. (1992). Application of the new production philosophy to construction. Stanford University (Technical Report No. 72, Center for Integrated Facility Engineering, Department of Civil Engineering). Stanford, CA.

Koskela, L. (2000). An exploration towards a production theory and its application to construction. VTT Technical Research Centre of Finland.

Koskela, L. (2003). "Is structural change the primary solution to the problems of construction?" Building Research & Information, 31(2), 85–96.

Koskela, L. and Howell, G. (2002). "The theory of project management: Explanation to novel methods." Proceedings 10th Annual Conference on Lean Construction, IGLC-10.

Koskela, L. J. (2004). "Moving on-beyond lean thinking." Lean Construction Journal, 1(1), 24–37.

LAI, M. P. D. T. (1998). "Identifying the Value Stream in Product Development". Cambridge, MA, Massachusetts Institute of Technology.

Lai, X. and Gershenson, J. K. (2008). "Representation of similarity and dependency for assembly modularity." The International Journal of Advanced Manufacturing Technology, 37(7-8), 803–827.

Langdon, D. (2004). "Costing green: A comprehensive cost Database and Budgeting Methodology." Lisa Fay Matthiesen and Peter Morris. July.

Lessing, J. (2006). "Industrialised house-building." Licentiate thesis, Division of Design Methodology, Lund Institute of Technology.

Liker J.K. and Morgan J. (2006). "The Toyota product development system: integrating people, process and technology. " Productivity Press.

Linnik, M., Berghede, K. and Ballard, G. (2013). "An experiment in takt time planning applied to non-repetitive work." Proc. 21 st Annual Conf. of the Int'l. Group for Lean Constr.

Lu, W., Erikshammar, J. and Olofsson, T. (2011). “Integrating Design and Production: A Component-Oriented Framework for Industrialized Housing.” Proceedings of the 2011 eg-ice Workshop. – Twente: University of Twente.

Malen, D. E. and Hancock, W. M. (1995). “Engineering for the customer: combining preference and physical system models. Part II—application.” Journal of Engeering Design, 6(4), 329–341.

Malmgren, L., Jensen, P. and Olofsson, T. (2010). “Product modeling of configurable building systems – a case study.” Journal of Information Technology in Construction, 15, 354–68.

Manavazhi, M. R. and Xunzhi, Z. (2001). “Productivity oriented analysis of design revisions.” Construction Management & Economics, 19(4), 379–391.

Martin, M. V. and Ishii, K. (2002). “Design for variety: developing standardized and modularized product platform architectures.” Research in Engineering Design, 13(4), 213–235.

Mikkelsen, H. and Riis, J.O., (2003), Grunnbog i prosjektledelse [Textbook of Project Management], Prodevo ApS, Rungsted

Mohamad, A., Hickethier, G., Hovestadt, V. and Gehbauer, F. (2013). "Use of modularization and standardization as a strategy to reduce component variety in On-off Projects" Proceedings of 21th International Group for Lean Construction Conference., Fortaleza, Brazil, 289-298

Ohon, T. (1988). "The Toyota Production System: Beyond Large-Scale Production". Productivity Press. OR.

Olsson, N. O. (2008). “External and internal flexibility–aligning projects with the business strategy and executing projects efficiently.” International Journal of Project Organisation and Management, 1(1), 47–64.

Oppenheim, W. (2011). “Lean for system engineering with lean enablers for system engineering ” Wiley-Blackwell, New Jersey.

Pandremenos, J., Paralikas, J., Salonitis, K. and Chryssolouris, G. (2009). "Modularity concepts for the automotive industry: a critical review." CIRP Journal of Manufacturing Science and Technology, 1(3), 148–152.

Parshall, S., Pena, W. and Kelly, K. (1987). Problem Seeking, An Architectural Programming Primer. AIA Press Washington, DC.

Pati, D., Harvey, T. and Cason, C. (2008). "Inpatient unit flexibility design characteristics of a successful flexible unit." Environment and Behavior, 40(2), 205–232.

Peña. (1987). "Problem seeking, An Architectural Programming pimer." AIA Press. Third Edition, Washington.

Poppendieck, M. and Poppendieck, T. (2003). "Lean software development: an agile toolkit. " Addison-Wesley Professional.

Radosavljević, M. and Horner, R. M. W. (2002). "The evidence of complex variability in construction labour productivity." Construction Management & Economics, 20(1), 3–12.

Rossi, M., Taisch, M. and Terzi, S. (2012). "Lean product development: A five-steps methodology for continuous improvement." Engineering, Technology and Innovation (ICE), 2012 18th International ICE Conference on, IEEE, 1–10.

Sacks, R. and Goldin, M. (2007). "Lean management model for construction of high-rise apartment buildings." Journal of construction engineering and Management, 133(5), 374–384.

Sacks, R., Koskela, L., Dave, B. A. and Owen, R. (2010a). "Interaction of lean and building information modeling in construction." *Journal of construction engineering and management*, 136(9), 968–980.

Sacks, R., Radosavljevic, M. and Barak, R. (2010b). "Requirements for building information modeling based lean production management systems for construction." *Automation in construction*, 19(5), 641–655.

Sako, M. and Murray, F. (1999). "Modules in design, production and use: implications for the global auto industry." IMVP Annual Sponsors Meeting, Citeseer.

Saxon, R. (2005). "Be Valuable: A guide to creating value in the built environment." Constructing Excellence in the Built Environment, London. 50 p

Schmidt III, R., Mohyuddin, S., Austin, S., Gibb, A. and others. (2008). "Using DSM to redefine buildings for adaptability." DSM 2008: Proceedings of the 10th International DSM Conference, Stockholm, Sweden, 11.-12.11. 2008.

Schregenberger, Johann W. Methodenbewusstes Problemlösen. (1980). ein Beitrag zur Ausbildung von Konstrukteuren, Beratern und Führungskräften. http://dx.doi.org/10.3929/ethz-a-000215324)

Seppänen, O. and others. (2009). "Empirical research on the success of production control in building construction projects." PhD dissertation, Faculty of Engineering and Architecture, Helsinki University of Technology.

Shamsuzzoha, A. H. M. and Helo, P. T. (2012). "Development of modular product architecture through information management." VINE, 42(2), 172–190.

Siyam, G. I., Wynn, D. C. and Clarkson, P. J. (2015). "Review of Value and Lean in Complex Product Development." Systems Engineering.

Stone, R. B., Wood, K. L. & Crawford, R. H. (2000). "A heuristic method for identifying modules for product architecture. Design Studies". 21 (1), 5-31.

Swaminathan, J. M. (2001). "Using Standardized Operations." California Management Review, 43(3), 125.

Tatum, C. B. and Korman, T. (1999). "MEP Coordination in Building and Industrial Projects." CIFE Work.

Tatum, C. B. and Korman, T. (2000). "Coordinating building systems: process and knowledge." Journal of architectural engineering, 6(4), 116–121.

Thomke, S. H. (1997). "The role of flexibility in the development of new products: An empirical study." Research policy, 26(1), 105–119.

Thurston, D. L. (1991). "A formal method for subjective design evaluation with multiple attributes." Research in engineering design, 3(2), 105–122.

Tiwari, S. and Sarathy, P. (2012). "Pull planning as a mechanism to deliver constructible design." IGLC 2012: 20th Conference of the International Group for Lean Construction, San Diego, CA, 17–22.

Tommelein, I. D. (2006). "Process benefits from use of standard products-simulation experiments using the pipe spool model." Proceedings of the 14th Annual Conference of the International Group for Lean Construction, 177–189.

Ulrich, K. (1995). "The role of product architecture in the manufacturing firm." Research policy, 24(3), 419–440.

Ulrich, K. a. S. E. (1995). "Product Design and Development." New York, NY, McGraw-Hill, Inc.

Ulrich, K. T. & Eppinger, S. D. (2004). Product Design and Development. McGraw-Hill. 3rd edition. ISBN 0-07-247146-8

Ulrich, K. T., Eppinger, S. D., (2008). "Product Design and Development", McGraw-Hill Eduction Asia, Singapore.

Upton, D. (1994). "The management of manufacturing flexibility." California management review, 36(2), 72–89.

Veenstra, V. S., Halman, J. I. and Voordijk, J. T. (2006). "A methodology for developing product platforms in the specific setting of the housebuilding industry." Research in engineering design, 17(3), 157–173.

Velamati, S. (2012). "Feasibility, benefits and challenges of modular construction in high rise development in the United States: a developer's perspective." Massachusetts Institute of Technology.

Walton, M. (1999). "Strategies for lean product development."

Ward, A., Liker, J. K., Cristiano, J. J. and Sobek, D. K. (1995). "The Second Toyota Paradox: How Delaying Decisions Can Make Better Cars Faster." Long Range Planning, 28(4), 129–129.

Ward, A. C. and Sobek, D. K., II. (2014). "*Lean Product and Process Development.* " Lean Enterprises Inst Inc, Cambridge, MA.

Womack, J. and Jones, D., Roos D. (1990). The Machine That Changed the World. Free Press.

Womack, J. a. D. T. J. (1996). Lean Thinking. New York, NY, Simon & Schuster.

Womack, J. and Jones, D. (2003). "Lean Thinking: Banish Waste and Create Wealth in your Corporation". Revised and updated edition. Simon & Schuster UK Ltd, London.

Zokaei, A. K. and Simons, D. W. (2006). "Value chain analysis in consumer focus improvement: a case study of the UK red meat industry." International Journal of Logistics Management, The, 17(2), 141–162.

FORSCHUNGSBERICHTE DES INSTITUTS FÜR TECHNOLOGIE UND MANAGEMENT IM BAUBETRIEB

Heft 1–62 institutsintern verlegt

Heft 11 **EMIL MASSINGER** 1976
Das rheologische Verhalten von lockeren Erdstoffgemischen

Heft 12 **KAWUS SCHAYEGAN** 1975
Einfluß von Bodenkonsistenz und Reifeninnendruck auf die fahrdynamischen Grundwerte von EM-Reifen

Heft 13 **CURT HEUMANN** 1975
Dynamische Einflüsse bei der Schnittkraftbestimmung in standfesten Böden

Heft 14 **HANS-JOSEF KRÄMER** 1976
Untersuchung der bearbeitungstechnischen Bodenkennwerte mit schwerem Ramm-Druck-Sondiergerät zur Beurteilung des Maschineneinsatzes im Erdbau

Heft 15 **FRIEDRICH ULBRICHT** 1977
Baggerkraft bei Eimerkettenschwimmbaggern - Untersuchungen zur Einsatzdimensionierung

Heft 16 **BERTOLD KETTERER** 1977
Einfluß der Geschwindigkeit auf den Schneidvorgang in rolligen Böden
- vergriffen -

Heft 17 **JOACHIM HORNUNG/THOMAS TRÜMPER** 1977
Entwicklungstendenzen lärmarmer Tiefbauverfahren für den innerstädtischen Einsatz

Heft 18 **JOACHIM HORNUNG** 1978
Geometrisch bedingte Einflüsse auf den Vorgang des maschinellen Reißens von Fels - untersucht an Modellen

Heft 19 **THOMAS TRÜMPER** 1978
Einsatzoptimierung von Tunnelvortriebsmaschinen

Heft 20 **GÜNTHER GUTH** 1978
Optimierung von Bauverfahren - dargestellt an Beispielen aus dem Seehafenbau

Heft 21 **KLAUS LAUFER** 1978
Gesetzmäßigkeiten in der Mechanik des drehenden Bohrens im Grenzbereich zwischen Locker- und Festgestein
- vergriffen -

Heft 22	**URS BRUNNER** Submarines Bauen - Entwicklung eines Bausystems für den Einsatz auf dem Meeresboden **- vergriffen -**	1979
Heft 23	**VOLKER SCHULER** Drehendes Bohren in Lockergestein - Gesetzmäßigkeiten und Nutzanwendung **- vergriffen -**	1979
Heft 24	**CHRISTIAN BENOIT** Die Systemtechnik der Unterwasserbaustelle im Offshore-Bereich	1980
Heft 25	**BERNHARD WÜST** Verbesserung der Umweltfreundlichkeit von Maschinen, insbesondere von Baumaschinen-Antrieben	1980
Heft 26	**HANS-JOSEF KRÄMER** Geräteseitige Einflußparameter bei Ramm- und Drucksondierungen und ihre Auswirkungen auf den Eindringwiderstand	1981
Heft 27	**BERTOLD KETTERER** Modelluntersuchungen zur Prognose von Schneid- und Planierkräften im Erdbau	1981
Heft 28	**HARALD BEITZEL** Gesetzmäßigkeiten zur Optimierung von Betonmischern	1981
Heft 29	**BERNHARD WÜST** Einfluß der Baustellenarbeit auf die Lebensdauer von Turmdrehkranen	1982
Heft 30	**HANS PINNOW** Einsatz großer Baumaschinen und bisher nicht erfaßter Sonderbauformen in lärmempfindlichen Gebieten	1982
Heft 31	**WALTER BAUMGÄRTNER** Traktionsoptimierung von EM-Reifen in Abhängigkeit von Profilierung und Innendruck	1982
Heft 32	**KARLHEINZ HILLENBRAND** Wechselwirkung zwischen Beton und Vibration bei der Herstellung von Stahlbetonrohren im Gleitverfahren	1983
Heft 33	**CHRISTIAN BENOIT** Ermittlung der Antriebsleistung bei Unterwasserschaufelrädern	1985

Heft 34 **NORBERT WARDECKI** 1986
Strömungsverhalten im Boden-/Werkzeugsystem

Heft 35 **CHRISTIAN BENOIT** 1986
Meeresbergbau - Bestimmung der erforderlichen
Antriebskraft von Unterwasserbaggern

Heft 36 **ROLF VICTOR SCHMÖGER** 1987
Automatisierung des Füllvorgangs bei Scrapern

Heft 37 **ALEXANDER L. MAY** 1987
Analyse der dreidimensionalen
Schnittverhältnissen beim Schaufelradbagger

Heft 38 **MICHAEL HELD** 1989
Hubschraubereinsatz im Baubetrieb

Heft 39 **GUNTER SCHLICK** 1989
Adhäsion im Boden-Werkzeug-System

Heft 40 **FRANZ SAUTER** 1991
Optimierungskriterien für das Unterwasser-
schaufelrad (UWS) mittels Modellsimulation
- vergriffen -

Heft 41 **STEFAN BERETITSCH** 1992
Kräftespiel im System Schneidwerkzeug-Boden

Heft 42 **HEINRICH SCHLICK** 1994
Belastungs- und Fließverhältnisse in Silos
mit zentralen Einbauten und Räumarmaustrag

Heft 43 **GÜNTHER DÖRFLER** 1995
Untersuchungen der Fahrwerkbodeninteraktion
zur Gestaltung von Raupenfahrzeugen für die
Befahrung weicher Tiefseeböden

Heft 44 **AXEL OLEFF** 1996
Auslegung von Stellelementen für
Schwingungserregerzellen mit geregelter
Parameterverstellung und adaptive
Regelungskonzepte für den Vibrationsrammprozeß

Heft 45 **KUNIBERT LENNERTS** 1997
Stand der Forschung auf den Gebieten der
Facility- und Baustellen-Layoutplanung

Heft 46 **KUNIBERT LENNERTS** 1997
Ein hybrides, objektorientiertes System zur
Planung optimierter Baustellen-Layouts

Heft 47	**UWE RICKERS** Modellbasiertes Ressourcenmanagement für die Rettungsphase in Erdbebengebieten	1998
Heft 48	**ULRICH-PETER REHM** „Ermittlung des Antriebsdrehmomentes von Räumarmen in Silos mit Einbaukörper und kohäsivem Schüttgut"	1998
Heft 49	**DIRK REUSCH** Modellierung, Parameterschätzung und automatische Regelung mit Erschütterungsbegrenzung für das langsame Vibrationsrammen	2001
Heft 50	**FRANZ DIEMAND** Strategisches und operatives Controlling im Bauunternehmen	2001
Heft 51	**KARSTEN SCHÖNBERGER** Entwicklung eines Workflow-Management- Systems zur Steuerung von Bauprozessen in Handwerkernetzwerken	2002
Heft 52	**CHRISTIAN MEYSENBURG** Ermittlung von Grundlagen für das Controlling in öffentlichen Bauverwaltungen	2002
Heft 53	**MATTHIAS BURCHARD** Grundlagen der Wettbewerbsvorteile globaler Baumärkte und Entwicklung eines Marketing Decision Support Systems (MDSS) zur Unternehmensplanung	2002
Heft 54	**JAROSŁAW JURASZ** Geometric Modelling for Computer Integrated Road Construction (Geometrische Modellierung für den rechnerintegrierten Straßenbau)	2003
Heft 55	**SASCHA GENTES** Optimierung von Standardbaumaschinen zur Rettung Verschütteter	2003
Heft 56	**GERHARD W. SCHMIDT** Informationsmanagement und Transformationsaufwand im Gebäudemanagement	2003
Heft 57	**KARL LUDWIG KLEY** Positionierungslösung für Straßenwalzen - Grundlage für eine kontinuierliche Qualitätskontrolle und Dokumentation der Verdichtungsarbeit im Asphaltbau	2004

Heft 58	**JOCHEN WENDEBAUM** Nutzung der Kerntemperaturvorhersage zur Verdichtung von Asphaltmischgut im Straßenbau	2004
Heft 59	**FRANK FIEDRICH** Ein High-Level-Architecture-basiertes Multiagentensystem zur Ressourcenoptimierung nach Starkbeben	2004
Heft 60	**JOACHIM DEDEKE** Rechnergestützte Simulation von Bauproduktions-prozessen zur Optimierung, Bewertung und Steuerung von Bauplanung und Bauausführung	2005
Heft 61	**MICHAEL OTT** Fertigungssystem Baustelle - Ein Kennzahlensystem zur Analyse und Bewertung der Produktivität von Prozessen	2007
Heft 62	**JOCHEN ABEL** Ein produktorientiertes Verrechnungssystem für Leistungen des Facility Management im Krankenhaus	2007

HEFT 63–68 BEI KIT SCIENTIFIC PUBLISHING KARLSRUHE VERLEGT, ISSN 1868-5951

Heft 63	**JÜRGEN KIRSCH** Organisation der Bauproduktion nach dem Vorbild industrieller Produktionssysteme – Entwicklung eines Gestaltungsmodells eines ganzheitlichen Produktionssystems für den Bauunternehmer	2009
Heft 64	**MARCO ZEIHER** Ein Entscheidungsunterstützungsmodell für den Rückbau massiver Betonstrukturen in kerntechnischen Anlagen	2009
Heft 65	**MARKUS SCHÖNIT** Online-Abschätzung der Rammguttragfähigkeit beim langsamen Vibrationsrammen in nichtbindigen Böden	2009
Heft 66	**JOHANNES KARL WESTERMANN** Betonbearbeitung mit hydraulischen Anbaufräsen	2009
Heft 67	**FABIAN KOHLBECKER** Projektbegleitendes Öko-Controlling Ein Beitrag zur ausgewogenen Bauprojektrealisierung beispielhaft dargestellt anhand von Tunnelbauprojekten	2010
Heft 68	**AILKE HEIDEMANN** Kooperative Projektabwicklung im Bauwesen unter der Berücksichtigung von Lean-Prinzipien - Entwicklung eines Lean-Projektabwicklungssystems: Internationale Untersuchungen im Hinblick auf die Umsetzung und Anwendbarkeit in Deutschland	2011

Die Bände sind unter www.ksp.kit.edu als PDF frei verfügbar oder als Druckausgabe bestellbar.

AB HEFT 69 BEI KIT SCIENTIFIC PUBLISHING KARLSRUHE UNTER DEM TITEL KARLSRUHER REIHE TECHNOLOGIE UND MANAGEMENT IM BAUBETRIEB VERLEGT, ISSN 2363-8222

Heft 69	**KIM KIRCHBACH** Anwendung von Lean-Prinzipien im Erdbau – Entwicklung eines Baustellenleitstands auf Basis von Virtual Reality	2015
Heft 70	**PATRICK KERN** Elastomerreibung und Kraftübertragung beim Abscheren von aktiv betriebenen Vakuumgreifern auf rauen Oberflächen	2017
Heft 71	**GERNOT HICKETHIER** Communication Structures in the Design Phase of Lean Project Delivery	2019
Heft 72	**AHLAM MOHAMAD** Managing the Potential of Modularization and Standardization of MEP Systems in Buildings. Guidelines for improvement based on lean principles	2019